STUDENT'S SOLUTIONS MANUAL

Elementary Algebra

W. Roy Fraser

Skyline College

Prepared by
Mark Serebransky
Camden County College
&
Stephen Taylor
Bucks County Community College

PWS PUBLISHING COMPANY

BOSTON

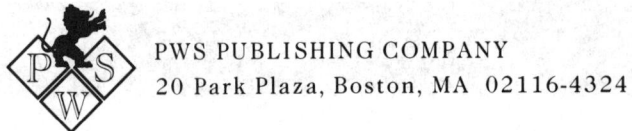

PWS PUBLISHING COMPANY
20 Park Plaza, Boston, MA 02116-4324

Copyright © 1994 by PWS Publishing Company.

All rights reserved. No part of this book may be reproduced, stored in a retrieval system, or transcribed in any form or by any means, electronic, mechanical, photocopying, or otherwise, without the prior written permission of the publisher, PWS Publishing Company.

PWS Publishing Company is a division of Wadsworth, Inc.

International Thomsom Publishing
The trademark ITP is used under license.

ISBN 0-534-93751-9

Printed and bound in the United States of America.
1 2 3 4 5 6 7 8 9 -- 99 98 97 96 95 94

Printer: Malloy Lithographing

Table of Contents

Chapter 1: The Real Number System .... Page 1

Chapter 2: Linear Equations and Inequalities .... Page 17

Chapter 3: Polynomials .... Page 49

Chapter 4: Factoring .... Page 64

Chapter 5: Rational Expressions .... Page 81

Chapter 6: The Cartesian Coordinate System .... Page 104

Chapter 7: Linear Systems .... Page 134

Chapter 8: Exponents and Radicals .... Page 159

Chapter 9: Quadratic Equations .... Page 179

Chapter 10: Additional Topics .... Page 200

Appendix A: Sets .... Page 214

Appendix B: Review of Fractions .... Page 216

PREFACE

This manual contains detailed solutions to the alternate odd-numbered exercises in the text *Elementary Algebra* by W. Roy Fraser. Extra attention has been paid to the more difficult exercises in order to improve the students' understanding of their solutions.

In many cases, especially in the case of word problems, the statement of the problem has been included so that the student does not have to refer back to the textbook when analyzing the solution.

We would appreciate any feedback concerning errors, solutions, corrections, solution style, or manual style. These and any other comments may be sent directly to us or in care of the publisher.

Mark Serebransky   Stephen Taylor
Camden County College   Bucks County Community College
P.O. Box 200   Swamp Road
Blackwood, NJ 08012   Newtown, PA 18940

We dedicate this book to our wives, NINA and DEBORAH, and thank them for their support and understanding.

Chapter 1: The Real Number System

Exercises 1.1

[1] A is equal to B is expressed as $A = B$.

[5] The set consisting of the 1st four natural numbers is $\{1,2,3,4\}$.

[9] {All the odd numbers between 8 and 16} = $\{9, 11, 13, 15\}$.

[13] {All the even primes} = $\{2\}$ since 2 is the only even prime number. For example, 4 is not prime because $4 = 2 \times 2$ and 6 is not prime because $6 = 2 \times 3$.

[17] The set of even numbers = $\{2, 4, 6, ...\}$ is infinite since it has an unlimited number of elements.

[21] The subsets of $\{9\}$ are $\{9\}$ and $\emptyset$. Note that $\emptyset$ is always a subset.

[25] The subsets of $\{6, 7, 8\}$ are $\{6, 7, 8\}$, $\{6, 7\}$, $\{6, 8\}$, $\{7, 8\}$, $\{6\}$, $\{7\}$, $\{8\}$, and $\emptyset$.

[29] The statement $\{x\} \subseteq \{x, y\}$ is true since each element of the 1st set (just x) is also an element of the 2nd set.

[33] The statement N = W is false because the two sets do not have exactly the same elements. Note that 0 is a whole number, but 0 is not a natural number.

[37] The prime numbers in $\{0, 1, 2, 3, 4, 5, 18, 23, 27, 28, 29, 30\}$ are 2, 3, 5, 23, and 29. These are the only numbers in the set which can not be broken down into a product of 2 smaller natural numbers.

[41] $176 + 2146 + 10119 = 12{,}441$

[45] $942 \times 630 = 593{,}460$

Exercises 1.1

[49] $\dfrac{12{,}988}{34} = 382$

[53] $82.615 + 127.035 = 209.650 = 209.65$

Exercises 1.2

[1] The graph of the set of numbers $\{0, 2, 4, 6, 8\}$ on the number line is

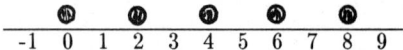

[5] The statement "7 is greater than 2" is written $7 > 2$.

[9] The statement "12 is less than 14" is written $12 < 14$.

[13] The answer to this question is $9 \geq 9$. Note that both $9 \leq 9$ and $9 \geq 9$ are true.

[17] If we wish to reverse the inequality of ".2 < 1.3" which is true and obtain a true statement then we must write $1.3 > .2$.

[21] $6(8) = 48$

[25] $\dfrac{8}{0}$ is undefined. Any division by 0 is undefined.

[29] $8 - (5 - 3) = 8 - 2 = 6$

[33] If $x = 4$ and $y = 3$ then $x(y+5) = 4(3+5) = 4(8) = 32$.

[37] If $x = 4$ and $y = 3$ then $\dfrac{x+8}{y-3} = \dfrac{4+8}{3-3} = \dfrac{12}{0}$ which is undefined.

[41] $38^1 = 38$

[45] $(1.23)^3 = (1.23)(1.23)(1.23) = 1.860867 \approx 1.86$

[49] $7 \cdot 7 \cdot 7 = 7^3$

[53] $b \cdot b \cdot b \cdot b = b^4$

[57] If $x = 2$ then $x^3 = 2^3 = 8$.

[61] The statement "The sum of a number and 4" is expressed as $x + 4$, where x represents the unknown number.

Exercises 1.2

65 The statement "The quotient of a number and 20" is expressed as $\frac{x}{20}$.

69 The statement "4 more than a number" is expressed as $x + 4$.

73 a) If $x = 2.3$ then $15.1 + x = 15.1 + 2.3 = 17.4$.
b) If $x = 3.15$ then $15.1 + x = 15.1 + 3.15 = 18.25$.

77 If $a = 18$, $b = 12.16$, and $c = 13.25$ then $(ab) - c = (18 \times 12.16) - 13.25 = 218.88 - 13.25 = 205.63$.

Exercises 1.3

1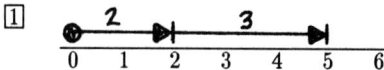

5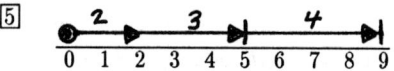

9 $6 + (7 + 8) = (6 + 7) + 8$ illustrates the Associative Property for Addition. Note that the order of the numbers on each side is the same, but the grouping on each side is different.

13 $(6 \cdot 8) \cdot 7 = 7 \cdot (6 \cdot 8)$ illustrates the Commutative Property for Multiplication. Note that this does not illustrate the associative property for multiplication because the parentheses on each side of the equation are around the $6 \cdot 8$. It's the order of the multiplication that has been changed.

17 $1 + (2 + 3) = (2 + 3) + 1$ illustrates the Commutative Property of Addition. Note that the grouping has not been changed.

21 $0^{11} = 0$ since $0^n = 0$ for all $n \neq 0$.

25 $\left(3 + \frac{7}{11}\right) + \frac{4}{11} = 3 + \left(\frac{7}{11} + \frac{4}{11}\right) = 3 + \frac{11}{11} = 4$.

Exercises 1.3

29 $2 + \frac{2}{3} = 2 \cdot \frac{3}{3} + \frac{2}{3} = \frac{2}{1} \cdot \frac{3}{3} + \frac{2}{3} = \frac{2 \cdot 3}{1 \cdot 3} + \frac{2}{3} = \frac{6}{3} + \frac{2}{3} = \frac{6+2}{3} = \frac{8}{3}$.

33 $\frac{8}{0} \cdot 0$ is undefined because $\frac{8}{0}$ is undefined. Any division by 0 is undefined.

37 $(87 \cdot 394)(6-6) = (87 \cdot 394)(0) = 0$. **41** $8(4x) = (8 \cdot 4)x = 32x$.

45 $(x+4) + (6+y) = x + [4 + (6+y)] = x + [(4+6) + y] = x + (10+y) = x + (y+10) = x + y + 10$.

49 $(2x)(4y)(3z) = (2 \cdot 4 \cdot 3)(xyz) = [(2 \cdot 4) \cdot 3](xyz) = (8 \cdot 3)(xyz) = 24xyz$.

53 a) The operation $a*b = (3a) + b^2$ is not commutative. To illustrate this, let $a = 2$ and $b = 3$. Then $2*3 = (3 \cdot 2) + 3^2 = 6 + 9 = 15$; but $3*2 = (3 \cdot 3) + 2^2 = 9 + 4 = 13$. Since $2*3 \neq 3*2$, the operation is not commutative.

b) The operation $a*b = (3a) + b^2$ is not associative. To illustrate this, let $a = 1$, $b = 2$, and $c = 3$. Then $(1*2)*3 = [(3 \cdot 1) + 2^2]*3 = (3+4)*3 = 7*3 = (3 \cdot 7) + 3^2 = 21 + 9 = 30$.

But $1*(2*3) = 1*(3 \cdot 2 + 3^2) = 1*(6+9) = 1*15 = 3 \cdot 1 + 15^2 = 3 + 225 = 228$.

Since $(1*2)*3 \neq 1*(2*3)$, the operation is not associative.

Review Problems

61 $\frac{8+2}{9-4} = \frac{10}{5} = 2$.

Exercises 1.4

1 $3 + 2 \cdot 6 = 3 + 12 = 15$. **5** $5(3 \cdot 7) = 5(21) = 105$.

Exercises 1.4

9 To evaluate $20 \div 5 \cdot 2$, we work from left to right doing any multiplication or division, whichever comes first. Therefore we first do $20 \div 5 = 4$, and then multiply this by 2. Therefore $20 \div 5 \cdot 2 = 4 \cdot 2 = 8$.

13 To evaluate $3^{11} + 20 \div 4$, we first do $3^{11} = 177{,}147$. Then we do $20 \div 4 = 5$. Therefore $3^{11} + 20 \div 4 = 177{,}147 + 5 = 177{,}152$. Note that the addition is done last.

17 To evaluate $2 + 3[(8-2)+4]$, we first evaluate what is in the parentheses which is $8 - 2 = 6$. Therefore $2 + 3[(8-2)+4] = 2 + 3[6+4]$. Then we evaluate what is in the brackets. $2 + 3(6+4) = 2 + 3(10) = 2 + 30 = 32$. Note we multiply the 3 times the 10 before we do the addition.

21 To evaluate $\frac{16 - 2 \cdot 3}{4 + 3 - 2}$, simplify the numerator and denominator separately. $16 - 2 \cdot 3 = 16 - 6 = 10$, and $4 + 3 - 2 = 5$. Therefore $\frac{16 - 2 \cdot 3}{4 + 3 - 2} = \frac{10}{5} = 2$.

25 If $x = 5$ and $y = 3$ then $2x + 4y = 2(5) + 4(3) = 10 + 12 = 22$.

29 If $x = 5$ and $y = 3$ then $x^2 + 2xy + y^2 = 5^2 + 2(5)(3) + 3^2 = 25 + 30 + 9 = 64$.

33 If $x = 5$ and $y = 3$ then $4x \div 2 - y = 4 \cdot 5 \div 2 - 3$. To simplify we first do the multiplication or division from left to right, whichever comes 1st, and do the subtraction last. Therefore $4 \cdot 5 \div 2 - 3 = 20 \div 2 - 3 = 10 - 3 = 7$.

37 If y denotes the unknown number then "8 subtracted from twice a number" is $2y - 8$.

41 The quotient of 7 times a number and 5 is expressed as $\frac{7y}{5}$.

45 To simplify $13 \cdot 1002$, write 1002 as $1000 + 2$. Then $13 \cdot 1002 = 13(1000 + 2) = 13 \cdot 1000 + 13 \cdot 2 = 13{,}000 + 26 = 13{,}026$.

Exercises 1.4

49 $8x + 3x = (8+3)x = 11x$.

53 $8 + (3+t) = (8+3) + t = 11 + t$.

57 $3p + 3 = 3p + 3 \cdot 1 = 3(p+1)$.

61 $6(a+b) = 6a + 6b$.

65 $3(2x + 4y) = 3(2x) + 3(4y) = (3 \cdot 2)x + (3 \cdot 4)y = 6x + 12y$.

69 If $x \neq 0$ then $\frac{0}{x} = 0$, not 2. If $x = 0$ then $\frac{0}{0}$ is undefined. Therefore $\frac{0}{x} = 2$ has no solution. It is a false equation.

73 The equation $5z = 5$ has the solution $z = 1$ since $5 \cdot 1 = 5$. It is a conditional equation.

77 From $A = \{0, 1, 2, 3, 4\}$, the only solution to $16 + z = 9z$ is $z = 2$.
$$16 + 2 = 9 \cdot 2 \text{ or } 18 = 18.$$

81 If x denotes the unknown number then the statement "The sum of 8 and a number is 12" is expressed as $x + 8 = 12$.

85 The statement "The product of a number and 4 is equal to 20" is expressed as $4x = 20$.

89 The statement "Twice the sum of a number and 3 is 10 more than the number" is expressed as $2(x + 3) = 10 + x$.

93 To evaluate $8.1 + 1.9(12)$, do the multiplication 1st. Then $8.1 + 1.9(12) = 8.1 + 22.8 = 30.9$.

97 To evaluate $(2.4)^2 - 3(1.04)$, do the subtraction last. Then $(2.4)^2 - 3(1.04) = 5.76 - 3.12 = 2.64$.

Exercises 1.4

Review Problems

105 The statement $4 > 2$ is true because 4 lies to the right of 2 on the number line.

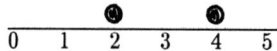

109 The statement $0 + 7 = 7$ illustrates the Additive Identity Property.

Exercises 1.5

1 A profit of $200 is the positive number 200.

5 20 seconds before lift off is the negative number -20.

9 $8 > 6$ since 8 lies to the right of 6 on the number line. The larger number lies to the right.

13 The statement $-6 \geq -5$ is false because -6 is not equal to -5 and -6 does not lie to the right of -5.

17 -8 is larger than -12 because -8 lies to the right of -12 on the number line.

21 The opposite of 8 is -8.

25 The opposite of -275 is 275.

29 If $a = -4$ then $-a = -(-4) = 4$.

33 The minus sign in $-(+4)$ denotes the opposite of the number $+4$.

Page 7

Exercises 1.5

37 The absolute value of -4 is $|-4| = -(-4) = 4$.

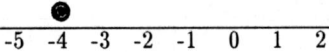

41 $|-8| = -(-8) = 8$.

45 If n denotes the number then the statement "The product of -3 and a number is 12" is expressed as $-3n = 12$.

49 The statement "-20 is 5 more than a number" is expressed as $-20 = n + 5$.

53 $\frac{0}{-8} = 0$ because $\frac{0}{a} = 0$ if $a \neq 0$.

57 $-12 + 0 = -12$ because of the additive identity property.

61 $-(+6) = -6$.

65 $3 + 2|-10| = 3 + 2 \cdot 10 = 3 + 20 = 23$.

69 $\frac{1}{2}[2(-9)] = \left(\frac{1}{2} \cdot 2\right)(-9) = 1(-9) = -9$.

73 $-\left[-(+5)\right] = -[-5] = 5$.

77 $|-6| < |10|$ since $|-6| = 6, |10| = 10$, and $6 < 10$.

81 $-x$ is **sometimes** a negative number. Note that if x is a negative number then $-x$ is a positive number. Therefore $-x$ is sometimes a positive number.

85 The set of negative integers is $\{\ldots -4, -3, -2, -1\}$. The largest negative integer is the number in the set which is furthest to the right. Therefore the largest negative integer is -1.

Review Problems

93 If $A = \{1, 2, 3\}$ then the subsets of A are $\{1, 2, 3\}, \{1, 2\}, \{1, 3\}, \{2, 3\}, \{1\}, \{2\}, \{3\}$, and $\emptyset$.

Exercises 1.6

[1]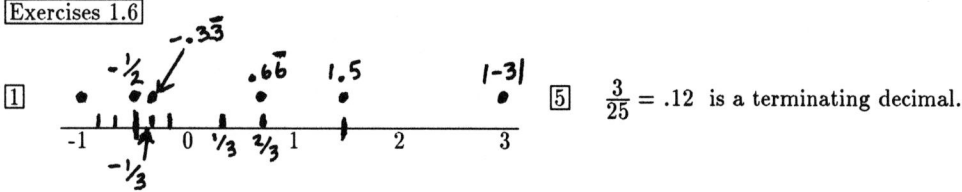

[5] $\frac{3}{25} = .12$ is a terminating decimal.

[9] $-\frac{201}{333} = -.60\overline{3603}$ is a repeating decimal.

[13] The rational numbers in the set A = $\{0, -17, -1.75, \pi, 8, 0.3\overline{3}, 0.31323334353637...\}$ are the numbers which have finite decimal expansions or infinite repeating decimal expansions. Therefore the rational numbers in A are $0, -17, -1.75, 8,$ and $0.3\overline{3}$.

[17] The integers in the set B = $\{2\frac{1}{2}, 7, -4, 0, 0.6\overline{6}, \frac{20}{4}, \sqrt{2}, 0.171819110111112...\}$ are $7, -4, 0,$ and $\frac{20}{4} = 5$.

[21] The statement "0 is both rational and irrational" is false. Since 0 is a rational number, it is not an irrational number. No number can be both rational and irrational, since a real number is an irrational number when it is not a rational number.

[25] The statement "The irrational numbers are a subset of the real numbers" is true because an irrational number is a real number which is not rational. Therefore all irrational numbers are real numbers.

[29] The statement "Some irrational numbers can be expressed as repeating decimals" is false because all repeating decimals are rational numbers.

[33] Since $8 + (-8) = 0$, it is a whole number, an integer, a rational number, and a real number. It is not a natural number, and it is not an irrational number.

[37] -8.25 is a rational number and a real number. It is not a natural number, a whole number, an integer, or an irrational number. It is a rational number because it is a terminating decimal.

Exercises 1.6

[41] $-\frac{7}{8}$ is a rational number and a real number. It is not a natural number, a whole number, an integer, or an irrational number.

Review Problems

[49]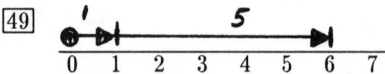

[53] $3^3 + 2|8-6| = 27 + 2|2| = 27 + 4 = 31$.

[57] 0 is a solution to $7y = 8y$ because $7(0) = 8(0)$ or $0 = 0$.

Exercises 1.7

[1]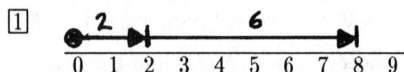

[5]

[9] $5 + (-3) = 2 + 3 + (-3) = 2 + [3 + (-3)] = 2 + 0 = 2$.

[13] Since $-20 = -12 + (-8)$ we can write
$-20 + 8 = -12 + (-8) + 8 = -12 + (-8+8) = -12 + 0 = -12$.

[17] $-8 + (-12) = -(8+12) = -20$.

[21] $-1566 + 3879 = 3879 - 1566 = 2313$.

[25] Since $30 > 15$, $15 + (-30) = -(30-15) = -15$.

[29] Since $-10 = -2 + (-8)$,
$-10 + 8 + 6 = -2 + (-8) + 8 + 6 = -2 + 0 + 6 = -2 + 6 = 6 - 2 = 4$.

[33] $1 + (-1) + 2 + (-2) = 0 + 0 = 0$.

[37] $5 \cdot 6 + (-40) = 30 + (-40) = -(40-30) = -10$.

Exercises 1.7

[41] $-18 + 20 \cdot 3 + (-1) = -18 + 60 - 1 = (-18 + 60) - 1 = 42 - 1 = 41$.

[45] $-|-8| + (-10) = -8 + (-10) = -(8 + 10) = -18$.

[49] $7 + 3|-8 + 10| = 7 + 3|2| = 7 + 3 \cdot 2 = 7 + 6 = 13$.

[53] $x = 10$ is not a solution to $2x + (-8) = -12$ because $2(10) + (-8) = 20 - 8 = 12$, not -12. $12 \neq -12$.

[57] $x = -5$ is a solution to $\frac{x+5}{x+6} = 0$ because $\frac{-5+5}{-5+6} = \frac{0}{1} = 0$.

[61] If Joe has $20 in his checking account and he cashes a check for $30 then his new balance is $20 - 30 = -\$10$.

[65] If Kim owed $250 on a charge card, made a payment of $120, and a week later charged a $40 sweater, then her final balance is $-250 + 120 - 40 = -250 + (120 - 40) = -250 + 80 = -(250 - 80) = -170$. Therefore she owed $170.

[69] $-14.21 + 6.07 = -(14.21 - 6.07) = -8.14$.

[73] $-\frac{2}{7} + \left(-\frac{3}{7}\right) = \frac{-2 + (-3)}{7} = \frac{-(2+3)}{7} = \frac{-5}{7} = -\frac{5}{7}$.

Review Problems

[81] $8 + 7(2) = 8 + 14 = 22$.

[85] $\frac{1}{6}(6x) = \left(\frac{1}{6} \cdot 6\right)x = 1x = x$.

[89] $4(6x + 3y) = 4 \cdot 6x + 4 \cdot 3y = 24x + 12y$.

Exercises 1.8

1. $2(4) = 4 + 4 = 8$.

5. $3(-4) = -12$.

9. $(-3)(-4) = 3 \cdot 4 = 12$.

13. a) The opposite of 8 is -8. b) The reciprocal of 8 is $\frac{1}{8}$. c) $|8| = 8$.

17. a) The opposite of $-\frac{2}{5}$ is $\frac{2}{5}$. b) The reciprocal of $-\frac{2}{5}$ is $-\frac{5}{2}$. c) $\left|-\frac{2}{5}\right| = \frac{2}{5}$.

21. a) The opposite of 2.06 is -2.06. b) The reciprocal is $\frac{1}{2.06} = .04854$. c) $|2.06| = 2.06$.

25. $(-9)\left(-\frac{1}{9}\right) = (9)\left(\frac{1}{9}\right) = 1$.

29. $-(-6) = 6$.

33. $(-6)^8 = 6^8 = 1{,}679{,}616$.

37. $-10 + (-3)^2 = -10 + 9 = -1$.

41. $-2[6 + (-8)] = -2[-2] = 4$.

45. $(-1.03)^4 = (1.03)^4 = 1.1255$.

49. $\dfrac{(-1)+(-3)^2}{(-4)^2+2(-8)} = \dfrac{-1+9}{16+(-16)} = \dfrac{8}{0}$ which is undefined.

53. If $x = 2$ and $y = -3$ then $-5x^2 + (-4)y^2 = -5(2)^2 + (-4)(-3)^2 = -5 \cdot 4 + (-4) \cdot 9 = -20 + (-36) = -(20 + 36) = -56$.

57. $-3(2x) = (-3 \cdot 2)x = -6x$.

61. $-2x + 2x = (-2 + 2)x = 0x = 0$.

65. $3[2x + (-4)] = 3(2x) + 3(-4) = (3 \cdot 2)x + (-12) = 6x - 12$.

69. $-5(-3z + 1) = (-5)(-3z) + (-5)(1) = (5 \cdot 3)z + (-5) = 15z - 5$.

73. If x denotes the number then "-2 times the sum of a number and 7" is expressed as $-2(x + 7)$.

Exercises 1.8

[77] The equation $4x = -28$ is a conditional equation. It has only one solution, $x = -7$.

[81] $-\frac{3}{4} + x = 0$ is a conditional equation with one solution, $x = \frac{3}{4}$.

[85] $-.3(-12)(.7) = [(-.3)(-12)](.7) = (3.6)(.7) = 2.52$.

Review Problems

[93] $\frac{20}{4} = 5$

[97] $\frac{8}{0}$ is undefined.

Exercises 1.9

[1] $\frac{12}{6} = 2$

[5] $\frac{0}{-8} = 0$

[9] $(-4013) \div 862 \approx -4.655$

[13] $-8 - (-4) = -8 + 4 = -(8-4) = -4$.

[17] $-5 - 6^2 = -5 - 36 = -41$.

[21] $3 - (2 \cdot 3 - 7) = 3 - (6 - 7) = 3 - (-1) = 3 + 1 = 4$.

[25] $(8 - 20) \div 2 = -12 \div 2 = -\left(\frac{12}{2}\right) = -6$.

[29] $\frac{-4^2 + 1}{2|-3|} = \frac{-(4^2) + 1}{2 \cdot 3} = \frac{-16 + 1}{2 \cdot 3} = \frac{-15}{6} = -\frac{\cancel{3} \cdot 5}{\cancel{3} \cdot 2} = -\frac{5}{2}$.

[33] $-\frac{9}{8} \div \frac{27}{10} = -\frac{9}{8} \cdot \frac{10}{27} = -\frac{\cancel{9}}{8} \cdot \frac{10}{\cancel{27}_3} = -\frac{10}{8 \cdot 3} = -\frac{5 \cdot \cancel{2}}{4 \cdot \cancel{2} \cdot 3} = -\frac{5}{12}$.

[37] If $x = -2$, $y = 3$, and $z = -5$ then $xy - z = (-2)(3) - (-5) = -6 + 5 = -1$.

Exercises 1.9

41 If $x = -2$, $y = 3$, and $z = -5$ then $\dfrac{x^3 + yz}{2xyz} = \dfrac{(-2)^3 + (3)(-5)}{2(-2)(3)(-5)} = \dfrac{-8 + (-15)}{60} = \dfrac{-23}{60}$.

45 $t = -3$ is not a solution to the equation $6 - 4t = 5t - 3$ because
$6 - 4(-3) = 6 - (-12) = 6 + 12 = 18$ but $5(-3) - 3 = -15 - 3 = -18$.
Since $18 \neq -18$, -3 is not a solution.

49 $a - 3 = a + (-3)$

53 $\dfrac{a-2}{6} = (a-2) \cdot \dfrac{1}{6}$

57 $(a - 10) - 14 = a + (-10) + (-14) = a + (-24) = a - 24$.

61 $3(2x - 4) = 3(2x) + 3(-4) = 6x + (-12) = 6x - 12$.

65 $-5(x + 6y - 3) = -5x + (-5)6y + (-5)(-3) = -5x - 30y + 15$.

69 $6x^2 - 6x = (6x) \cdot x - (6x) \cdot 1 = 6x(x - 1)$.

73 The statement "The difference of a number and -8 is 10" is expressed as $x - (-8) = 10$ or more simply, $x + 8 = 10$.

77 "When 3 is subtracted from the quotient of a number and 4, the result is -8" is expressed as $\dfrac{x}{4} - 3 = -8$.

81 If company A made a profit of \$60,000 while company B had a loss of \$40,000, then the difference between the profit and loss is $60,000 - (-40,000) = 60,000 + 40,000 = \$100,000$.

85 $\dfrac{123.7}{-16.9} = -\dfrac{123.7}{16.9} \approx -7.32$.

89 $1.035 - (-2.368) = 1.035 + 2.368 = 3.403$.

Chapter 1 Review Exercises and Test

Chapter Review Exercises

1. $2(-5) - 6 = -10 - 6 = -16$.

5. $(-2)^4 - 20 = 16 - 20 = -4$.

9. $5^2 - 6 + 8 \div (-2) = 25 - 6 + 8 \div (-2) = (25 - 6) - 4 = 19 - 4 = 15$.

13. $4y - 6x + xz = 4(-3) - 6(2) + 2(-5) = -12 - 12 - 10 = -34$.

17. $-10 + (x - 4) = -10 + x - 4 = -14 + x = x - 14$.

21. $-(8 - 6t) = -8 - (-6t) = -8 + 6t = 6t - 8$.

25. $(-6 + 6)x = 0 \cdot x$ illustrates the additive inverse property which states that $a + (-a) = 0$. This is property E.

29. a) The opposite of $\frac{2}{3}$ is $-\frac{2}{3}$.   b) Its reciprocal is $\frac{3}{2}$.   c) $\left|\frac{2}{3}\right| = \frac{2}{3}$.

33. $x = -2$ is not a solution to the equation $3x - (-4) = -10$ because $3(-2) - (-4) = -6 + 4 = -2$ and $-2 \neq -10$.

37. The rational numbers in the set $\{0, -3, 6, -3.4, .21\overline{21}, .2121121112...\}$ are $0, -3, 6, -3.4,$ and $.21\overline{21}$.

41. If $A = \{1, 2, 3\}$ the subsets are $\{1, 2, 3\}, \{1, 2\}, \{1, 3\}, \{2, 3\}, \{1\}, \{2\}, \{3\},$ and $\emptyset$.

45. The statement "If 1 less than a number is divided by 4, the result is 8" is expressed as $\frac{x-1}{4} = 8$.

49. If Fran has $40 in her checking account and she cashes a check for $45 then her new balance is $40 - 45 = -\$5$.

Chapter 1 Review Exercises and Test

Chapter Test

1. $-10-(-4) = -10+4 = -6$.

5. $(-4)^2 - (-8)(3) = 16 + 24 = 40$.

9. $-3 + (x-8) = -3 + x - 8 = -11 + x = x - 11$.

13. $-(4m - 3) = -4m - (-3) = -4m + 3$.

17. a) The opposite of $-\frac{2}{7}$ is $\frac{2}{7}$. b) Its reciprocal is $-\frac{7}{2}$. c) $\left|-\frac{2}{7}\right| = \frac{2}{7}$.

21. $x = -3$ is not a solution to the equation $6x - 1 = 8x + 2$ because
    $6(-3) - 1 = -18 - 1 = -19$ but $8(-3) + 2 = -24 + 2 = -22$.

25. The integers in the set $\{-4, 0, 4, 1 \div 0, 2.75, \frac{2}{3}, .4\overline{4}, \sqrt{2}\}$ are $-4, 0,$ and 4.

29. If the temperature at sunrise was $-30°F$ and it went up $26°F$ by noon, the new temperature is
    $-30 + 26 = -4°F$.

Chapter 2: Linear Equations and Inequalities

Exercises 2.1

1 The numerical coefficient of $23x$ is 23.

5 The numerical coefficient of $1.9xyz$ is 1.9.

9 $10x$ and x are like terms.

13 $6r$ and $-\frac{1}{5}r$ are like terms.

17 $4z + 2z + 6 = 12z$ is not true when $z = 4$ because $4(4) + 2(4) + 6 = 16 + 8 + 6 = 30$ which is not equal to $12(4) = 48$.

21 $-(4x - 3z) = -4x + 3z$ is true when $x = 3$ and $z = 4$.
$-(4 \cdot 3 - 3 \cdot 4) = -(12 - 12) = -0 = 0$ and $-4(3) + 3(4) = -12 + 12 = 0$.
Note that $-(4x - 3z) = -4x + 3z$ is true regardless of the values of x and z.

25 $4b + 10 + b = 4b + b + 10 = (4+1)b + 10 = 5b + 10$.

29 $4x + 3y - 4x + y = 4x - 4x + 3y + y = (4-4)x + (3+1)y = 0x + 4y = 4y$.

33 $-2z - 4z = (-2-4)z = -6z$.

37 $6t - 1 - 6t = 6t - 6t - 1 = (6-6)t - 1 = 0t - 1 = -1$.

41 $8a + 3a - 2a = (8+3-2)a = 9a$.

45 $3x^2 + 4y - 3x^2 - 7 = 3x^2 - 3x^2 + 4y - 7 = (3-3)x^2 + 4y - 7 = 0x^2 + 4y - 7 = 4y - 7$.

49 $-(x+8) + 5x = -x - 8 + 5x = -x + 5x - 8 = (-1+5)x - 8 = 4x - 8$.

53 $4(6x-2) + 6(6x-2) = (4 \cdot 6x - 4 \cdot 2) + (6 \cdot 6x - 6 \cdot 2) = (24x - 8) + (36x - 12) = 24x + 36x - 8 - 12 = 60x - 20$.

Exercises 2.1

[57] $\frac{1}{2}(6x-3)-\frac{1}{3}(6x-5)=\left[\frac{1}{2}\cdot 6x-\frac{1}{2}\cdot 3\right]+\left[-\frac{1}{3}\cdot 6x-\frac{1}{3}(-5)\right]=\left[3x-\frac{3}{2}\right]+\left[-2x+\frac{5}{3}\right]=$

$3x-2x-\frac{3}{2}+\frac{5}{3}=x-\frac{3}{2}\cdot\frac{3}{3}+\frac{5}{3}\cdot\frac{2}{2}=x-\frac{9}{6}+\frac{10}{6}=x+\frac{-9+10}{6}=x+\frac{1}{6}$.

[61] Let x be the number. Then "the sum of a number, five times the number, and 9 times the number is added to 20" is expressed as $x+5x+9x+20$.

[65] If x is the number then "twice the sum of a number and -8 is subtracted from the number" is given by $x-2(x-8)$. To derive this expression, do one bit at a time. For example, the sum of a number and -8 is given by $x+(-8)$ or $x-8$. Then twice the sum of a number and -8 is given by $2(x-8)$. Finally subtract $2(x-8)$ from x which gives $x-2(x-8)$.

[69] To simplify $3x+4[x-2(x+8)]$, 1st get rid of the parentheses and combine like terms.

$3x+4[x-2(x+8)]=3x+4[x-2x-16]=3x+4[-x-16]$. Now get rid of the brackets.

$3x+4[-x-16]=3x+4(-x)+4(-16)=3x-4x-64=-x-64$.

[73] $3(x-8)-2(4-3x)+5=3x-24-8+6x+5=9x-27$.

[77] $x-.25(x+3)-.21=x-.25x-.75-.21=(1.00x-.25x)-(.75+.21)=.75x-.96$.

Review Problems

[85] $(x+6)+(-6)=x$.

[89] $(-1)(-r)=(-1)(-1)r=1r=r$.

Exercises 2.2

[1] To solve $x+8=-17$ subtract 8 from both sides:

$x+8-8=-17-8$
$x+0=-25$
$x=-25$

Exercises 2.2

5 To solve $-4x = 24$ multiply both sides by $-\frac{1}{4}$ because $\left(-\frac{1}{4}\right)(-4) = 1$:

$$-\tfrac{1}{4}(-4x) = -\tfrac{1}{4}(24)$$
$$1x = -\tfrac{24}{4}$$
$$x = -6$$

9 To solve $6s = -16$, divide both sides by 6:

$$\tfrac{6s}{6} = -\tfrac{16}{6}$$
$$1s = -\tfrac{8 \cdot \cancel{2}}{3 \cdot \cancel{2}}$$
$$s = -\tfrac{8}{3}$$

13 To solve $3(6x+2) + 1 = 7 + 18x$ first remove the parentheses and combine like terms:

$$3 \cdot 6x + 3 \cdot 2 + 1 = 7 + 18x$$
$$18x + 6 + 1 = 7 + 18x$$
$$18x + 7 = 18x + 7 \;\Leftarrow\; \text{This equation is true for all real numbers because both sides}$$
$$\text{of the equation are identical.}$$

17 The equation $x - 1 = 6$ has the solution set $\{7\}$: $7 - 1 = 6$. However the equation $x = 5$ has the solution set $\{5\}$. Therefore $x - 1 = 6$ is not equivalent to $x = 5$.

21 The equations $-6x = -30$ and $x = 5$ are equivalent because both equations have the solution set $\{5\}$.

25 To solve $\tfrac{1}{2}r = 16$, multiply both sides by 2:

$2\left(\tfrac{1}{2}\right)r = 2(16)$ which gives the equivalent equation $1r = 32$ or $r = 32$.

Check: $\tfrac{1}{2}(32) = \tfrac{32}{2} = 16$.

29 $y - 7 = 20$ is equivalent to $y - 7 + 7 = 20 + 7$ or $y = 27$. Therefore the solution set is $\{27\}$.

33 $r - (-6) = 0$ is equivalent to $r + 6 = 0$. Then subtract 6 from both sides:

$r + 6 - 6 = 0 - 6$ or $r = -6$. Therefore the solution set is $\{-6\}$.

Exercises 2.2

37 To solve $z + 1.25 = .35$ subtract -1.25 from both sides:

$z + 1.25 - 1.25 = 0.35 - 1.25$ which implies that $z = -.9$. Therefore the solution set is $\{-.9\}$.

41 To solve $-3 = \frac{y}{10}$ multiply both sides of the equation by 10:

$-3(10) = \frac{y}{10} \cdot 10$ which implies that $-30 = y$. Therefore the solution set is $\{-30\}$.

45 The equation $0x = 0$ is true for all real numbers. Hence the solution set is $\Re$.

49 To solve $-\frac{7x}{8} = 2.2$ multiply both sides by $-\frac{8}{7}$:

$\left(-\frac{8}{7}\right)\left(-\frac{7}{8}\right)x = \left(-\frac{8}{7}\right)(2.2)$ which implies that $1x = -\frac{17.6}{7}$ or $x \approx -2.51$.

Therefore the solution set is $\{-2.51\}$.

53 To solve $13x - 4(3x - 2) = 12$ first remove the parentheses and combine like terms:

$13x - 12x + 8 = 12$
$x + 8 = 12$
$x + 8 - 8 = 12 - 8$
$x = 4$ The solution set is $\{4\}$.

57 $15 - 3(4x + 5) = 0$
$15 - 12x - 15 = 0$
$-12x = 0$
$x = \frac{0}{-12} = 0$ The solution set is $\{0\}$.

61 If the sum of a number and 7 is 2 then $x + 7 = 2$ where x represents the number. Then $x + 7 - 7 = 2 - 7$ or $x = -5$. Therefore the number is -5.

Check: $(-5) + 7 = 7 - 5 = 2$.

65 If the product of a number and 6 is -21 then $6x = -21$ where x represents the number. Then $\frac{6x}{6} = -\frac{21}{6}$ or $x = -\frac{21}{6} = -\frac{7 \cdot 3}{2 \cdot 3} = -\frac{7}{2}$. Therefore the number is $-\frac{7}{2}$.

Check: $6\left(-\frac{7}{2}\right) = -\frac{6 \cdot 7}{2} = -\frac{42}{2} = -21$.

Exercises 2.2

[69] If twice the sum of a number and 3 is equal to 6 more than twice a number then $2(x+3) = 2x + 6$ where x represents the number. Then $2x + 6 = 2x + 6$ which is true for all real numbers since both sides of the equation are the same. Therefore the number can be any real number.

[73] To solve $\frac{0}{r} + r + 8 = 8$ note that $r = 0$ can not be a solution since $\frac{0}{0}$ is undefined. If $r \neq 0$ then $\frac{0}{r} = 0$. Then the equation reduces to $r + 8 = 8$ or $r = 8 - 8 = 0$. However, as stated above, $r = 0$ can not be a solution. Therefore this equation has no solutions.

[77] To solve $13 - 3(2x + 1) + 2(4x - 1) = 3x + 8 - x$ remove all grouping symbols:
$13 - 6x - 3 + 8x - 2 = 3x + 8 - x$ which simplifies to
$8 + 2x = 8 + 2x$ which is satisfied by all real numbers since both sides of the equation are the same. Therefore the solution set is $\Re$.

[81] $2.76z = 1.04 - 8.43$ implies that $2.76z = -7.39$. Then divide both sides by 2.76:
$\frac{2.76z}{2.76} = -\frac{7.39}{2.76}$ or $z = -2.68$. Therefore the solution set is $\{-2.68\}$.

Review Problems

[89] $|9| = 9$ **[93]** $x = -8$ is a solution to $|x| = 8$ since $|-8| = 8$.

Exercises 2.3

[1] $2x + 8 = 16$ **[5]** $7x + 2 = -x - 14$
    $2x + 8 - 8 = 16 - 8$ $7x + 2 - 2 = -x - 14 - 2$
    $2x = 8$ $7x = -x - 16$
    $\frac{2x}{2} = \frac{8}{2}$ $7x + x = -x - 16 + x$
    $x = 4$ $8x = -16$ or $\frac{8x}{8} = -\frac{16}{8}$ or $x = -2$.

Exercises 2.3

9 $|8| = 8$

13 $-3|2 \cdot 9 - 4| = -3|18 - 4| = -3|14| = -3 \cdot 14 = -42$.

17 The equation $|r + 8| = -2$ has no solutions because $|r + 8|$ is greater than or equal to 0, regardless of the value of r. Therefore $|r + 8|$ can never equal a negative number -2.

21
$6x + 7 = 0$
$6x + 7 - 7 = 0 - 7$
$6x = -7$
$\frac{6x}{6} = \frac{-7}{6}$
$x = -\frac{7}{6}$. The solution set is $\{-\frac{7}{6}\}$.

25
$-1 = 8 - w$
$-1 - 8 = 8 - w - 8$
$-9 = -w$
$\frac{-9}{-1} = \frac{-w}{-1}$
$9 = w$. The solution set is $\{9\}$.

29
$x - 9 = 9 - x$
$x - 9 + 9 = 9 - x + 9$
$x = 18 - x$
$x + x = 18 - x + x$
$2x = 18$
$x = \frac{18}{2} = 9$. The solution set is $\{9\}$.

33
$10y + 4 = 9y - 1$
$10y + 4 - 9y = 9y - 1 - 9y$
$y + 4 = -1$
$y + 4 - 4 = -1 - 4$
$y = -5$. The solution set is $\{-5\}$.

37
$4x + 3 = 4x - 7$
$-4x + 4x + 3 = -4x + 4x - 7$
$3 = -7$. This is a false statement regardless of what number we put in for x. Therefore there are no solutions and the solution set is $\emptyset$.

41 $|x| = 4$ implies that $x = 4$ or $x = -4$. The solution set is $\{-4, 4\}$.

45
$3(2x + 5) = 6x + 5$
$6x + 15 = 6x + 5$
$-6x + 6x + 15 = -6x + 6x + 5$
$15 = 5$. This is a false statement and the equation has no solutions. The solution set is $\emptyset$.

49 $|2t - 1| = 7$ implies that
$2t - 1 = 7$ or $2t - 1 = -7$. Therefore
$2t = 8$ or $2t = -6$ which implies that
$t = \frac{8}{2} = 4$ or $t = -\frac{6}{2} = -3$.
Therefore the solution set is $\{-3, 4\}$.

Exercises 2.3

[53] $|9x| = -18$ has no solutions since $|9x|$ is never negative, regardless of the value of x. Therefore the solution set is $\emptyset$.

[57] $\left|\frac{k}{2} - 1\right| = 3$ implies that $\frac{k}{2} - 1 = 3$ or $\frac{k}{2} - 1 = -3$. Solve each equation separately:

$\frac{k}{2} - 1 + 1 = 3 + 1$ $\frac{k}{2} - 1 + 1 = -3 + 1$

$\frac{k}{2} = 4$ $\frac{k}{2} = -2$

$2 \cdot \frac{k}{2} = 2 \cdot 4$ $2 \cdot \frac{k}{2} = 2 \cdot (-2)$

$k = 8$ $k = -4$ The solution set is $\{-4, 8\}$.

[61] If one more than twice a number is 17 then $2x + 1 = 17$ where x represents the unknown number. Therefore $2x + 1 - 1 = 17 - 1$ or $2x = 16$ which implies that $x = \frac{16}{2} = 8$.
Check: $1 + 2 \cdot 8 = 1 + 16 = 17$.

[65] If 10 more than a number is equal to 5 less than twice the number then $x + 10 = 2x - 5$ where x is the unknown number. Therefore $x + 10 + 5 = 2x - 5 + 5$ or $x + 15 = 2x$.
Then add $-x$ to both sides which gives $-x + x + 15 = -x + 2x$ or $15 = x$.
Therefore the unknown number is 15.
Check: $10 + 15 = 25$ and $2 \cdot 15 - 5 = 30 - 5 = 25$.

[69] If x is the number and 3 increased by $\frac{3}{2}$ of the number is the same as twice the number then

$\frac{3}{2}x + 3 = 2x$ $\Leftarrow$ Now subtract $\frac{3}{2}x$ from both sides.

$-\frac{3}{2}x + \frac{3}{2}x + 3 = -\frac{3}{2}x + 2x$ $\Leftarrow$ $-\frac{3}{2}x + 2x = \left(-\frac{3}{2} + 2\right)x = \left(-\frac{3}{2} + \frac{4}{2}\right)x = \frac{1}{2}x$

$3 = \frac{1}{2}x$ $\Leftarrow$ Multiply both sides by 2

$2 \cdot 3 = 2 \cdot \frac{1}{2}x$ or $6 = x$. Therefore the number is 6.

[73] If the absolute value of 1 more than a number is 20 then $|x + 1| = 20$. Therefore $x + 1 = 20$ or $x + 1 = -20$. Solving each equation separately we get

$x + 1 - 1 = 20 - 1$ or $x + 1 - 1 = -20 - 1$

$x = 19$ or $x = -21$. Therefore the number can be -21 or 19.

Exercises 2.3

[77] One less than 7 times a number is given by $7x - 1$ where x is the number. The distance of $7x - 1$ from 0 is $|7x - 1|$. If this distance is 8 units then $|7x - 1| = 8$. Therefore $7x - 1 = 8$ or $7x - 1 = -8$ which implies that $7x = 9$ or $7x = -7$ which implies that $x = \frac{9}{7}$ or $x = -1$. Thus the number can be -1 or $\frac{9}{7}$.

[81] To solve $3 + 4|x| = 19$, subtract 3 from both sides to isolate the absolute value term:
$3 + 4|x| - 3 = 19 - 3$
$4|x| = 16$ ⇐ Now divide both sides by 4
$\frac{4|x|}{4} = \frac{16}{4}$ which simplifies to $|x| = 4$. Therefore $x = 4$ or -4.
Check: $3 + 4|4| = 3 + 4 \cdot 4 = 3 + 16 = 19$ and $3 + 4|-4| = 3 + 4 \cdot 4 = 3 + 16 = 19$.

[85] $1.4x + 1.9 = -8.6$ is equivalent to $1.4x = -8.6 - 1.9$ or $1.4x = -10.5$. Divide both sides by 1.4 and we get $x = \frac{-10.5}{1.4} = -7.5$.

[89] $|3x - 1| = x$ implies that $3x - 1 = x$ or $3x - 1 = -x$. Solving each equation, we get
$3x - 1 - x = x - x$ or $3x - 1 + x = -x + x$
$2x - 1 = 0$ or $4x - 1 = 0$ ⇐ Now add 1 to each side
$2x = 1$ or $4x = 1$ ⇐ Divide the 1st equation by 2 and the 2nd equation by 4
$x = \frac{1}{2}$ or $x = \frac{1}{4}$.

Review Problems

[97] $12x - 8 - 3x = (12x - 3x) - 8 = 9x - 8$.

[101] $12 - (8 - y) + 3 = 12 - 8 + y + 3 = y + 7$.

Exercises 2.4

1.
$3x - 1 = x + 7$
$-x + 3x - 1 = -x + x + 7$
$2x - 1 = 7$
$2x - 1 + 1 = 7 + 1$
$2x = 8$
$\frac{2x}{2} = \frac{8}{2}$ or $x = 4$.
Check: $3 \cdot 4 - 1 = 4 + 7$
$12 - 1 = 11$
$11 = 11$ ✓

5.
$1 - 12v + 9 = 3v - 4 - v$
$-12v + 10 = 2v - 4$
$-2v - 12v + 10 = -2v + 2v - 4$
$-14v + 10 = -4$
$-14v + 10 - 10 = -4 - 10$
$-14v = -14$
$\frac{-14v}{-14} = \frac{-14}{-14}$ or $v = 1$.
Check: $1 - 12 \cdot 1 + 9 = 3 \cdot 1 - 4 - 1$
$1 - 12 + 9 = 3 - 4 - 1$
$-2 = -2$ ✓

9.
$6(n + 2) = 4(n + 3)$
$6n + 12 = 4n + 12$
$6n + 12 - 12 = 4n + 12 - 12$
$6n = 4n$
$6n - 4n = 4n - 4n$
$2n = 0$
$\frac{2n}{2} = \frac{0}{2}$ or $n = \frac{0}{2} = 0$.
Check: $6(0 + 2) = 4(0 + 3)$
$6(2) = 4(3)$
$12 = 12$ ✓

13.
$5.2 - .2x = .6x - 2$
$2 + 5.2 - .2x = 2 + .6x - 2$
$7.2 - .2x = .6x$
$7.2 - .2x + .2x = .6x + .2x$
$7.2 = .8x$
$\frac{7.2}{.8} = \frac{.8x}{.8}$ or $x = \frac{7.2}{.8} = 9$.
Check: $5.2 - .2(9) = .6(9) - 2$
$5.2 - 1.8 = 5.4 - 2$
$3.4 = 3.4$ ✓

17.
$1 + 3(2x + 1) = 2(3x + 1)$ ⇐ Remove the grouping symbols
$1 + 6x + 3 = 6x + 2$ ⇐ Simplify
$6x + 4 = 6x + 2$ ⇐ Subtract $6x$ from both sides
$4 = 2$. ⇐ Thus we get a false statement regardless of the value of x.

Therefore the equation has no solutions and the solution set is ∅.

Exercises 2.4

21. $-(3p-1)-(6-4p)=2$ ⇐ Remove the grouping symbols
$\quad\quad -3p+1-6+4p=2$ ⇐ Simplify
$\quad\quad p-5=2$ ⇐ Add 5 to both sides
$\quad\quad p-5+5=2+5 \text{ or } p=7.$ Therefore the solution set is $\{7\}$.

25. $2[x+2(x-3)]=x-1$ ⇐ Remove the parentheses
$\quad\quad 2[x+2x-6]=x-1$ ⇐ Simplify
$\quad\quad 2[3x-6]=x-1$ ⇐ Remove the brackets
$\quad\quad 6x-12=x-1$ ⇐ Subtract x from both sides and add 12 to both sides
$\quad\quad 5x=11$ ⇐ Divide both sides by 5
$\quad\quad x=\frac{11}{5}.$ Therefore the solution set is $\{\frac{11}{5}\}$.

29. $(5-x)-1=5-(x-1)$ ⇐ Remove the parentheses
$\quad\quad 5-x-1=5-x+1$ ⇐ Simplify
$\quad\quad 4-x=6-x$ ⇐ Add x to both sides
$\quad\quad 4=6$ ⇐ This is a false statement. Therefore the equation has no solutions and the solution set is $\emptyset$.

33. $\frac{z}{5}+\frac{z}{6}=2(z-1)-2z$ ⇐ Remove the parentheses
$\quad\quad \frac{z}{5}+\frac{z}{6}=2z-2-2z$ ⇐ Simplify
$\quad\quad \frac{z}{5}+\frac{z}{6}=-2$ ⇐ Multiply both sides by $5\cdot 6=30$.
$\quad\quad 30\left(\frac{z}{5}+\frac{z}{6}\right)=-2\cdot 30$ ⇐ Use the distributive law to remove the parentheses
$\quad\quad 30\cdot\frac{z}{5}+30\cdot\frac{z}{6}=-60$ ⇐ Divide 5 into the 30 and divide 6 into the 30.
$\quad\quad 6z+5z=-60$ ⇐ Add like terms
$\quad\quad 11z=-60$ ⇐ Divide both sides by 11
$\quad\quad z=\frac{-60}{11}.$ Therefore the solution set is $\{-\frac{60}{11}\}$.

Exercises 2.4

[37] $6c + 4.5(20 - c) = 100$ ⇐ Remove the parentheses

 $6c + 90 - 4.5c = 100$ ⇐ Combine like terms and subtract 90 from both sides

 $1.5c = 10$ ⇐ Multiply both sides by 10

 $15c = 100$ ⇐ Divide both sides by 15

 $c = \dfrac{100}{15} = \dfrac{\not{5} \cdot 20}{\not{5} \cdot 3} = \dfrac{20}{3}$. Therefore the solution set is $\{\dfrac{20}{3}\}$.

[41] If a quantity and its seventh makes 19 then $x + \dfrac{1}{7}x = 19$ where x is the unknown quantity.
Then multiply both sides of the equation by 7:

 $7\left(x + \dfrac{1}{7}x\right) = 7 \cdot 19$ ⇐ Get rid of the parentheses using the distributive law

 $7x + x = 133$ or $8x = 133$ which implies that $x = \dfrac{133}{8}$.

 Check: $\dfrac{133}{8} + \dfrac{1}{7}\left(\dfrac{133}{8}\right) = \dfrac{133}{8} + \dfrac{7 \cdot 19}{7 \cdot 8} = \dfrac{133}{8} + \dfrac{19}{8} = \dfrac{152}{8} = 19$.

[45] The problem states that if ten times a number is subtracted from five times the number, then the result is 18 more than than the number. If x represents the number then ten times the number subtracted from five times the number is $5x - 10x$, and 18 more than the number is $x + 18$. Therefore $5x - 10x = x + 18$ which simplifies to $-5x = x + 18$. Subtract x from each side to obtain $-6x = 18$ or $x = \dfrac{18}{-6} = -3$. Therefore the number is -3.

 Check: $5(-3) - 10(-3) = -15 + 30 = 15$ and $-3 + 18 = 15$.

[49] Problem Statement: The sum of two numbers is 41. One less than twice the smaller number is equal to the larger number.

Let x be the smaller number. If L represents the larger number, then $x + L = 41$. Therefore $L = 41 - x$; so the larger number is $41 - x$.

One less than twice the smaller number is $2x - 1$ and since this is equal to the larger number we get the equation

 $2x - 1 = 41 - x$ ⇐ Add x to both sides and add 1 to both sides

 $3x = 42$ or $x = \dfrac{42}{3} = 14$.

Therefore the smaller number is 14 and the larger number is $41 - 14 = 27$.

Check: $14 + 27 = 41$ ✓ $2 \cdot 14 - 1 = 27$ or $28 - 1 = 27$ ✓

Exercises 2.4

[53] Problem Statement: The difference of two numbers is 17. The sum of 4 times the smaller number and twice the larger number is 52.

If x is the smaller number then the larger number minus x is 17 which implies that the larger number is $x + 17$ (or $L - x = 17$ implies $L = x + 17$).

The sum of 4 times the smaller number and twice the larger is $4x + 2(x + 17)$. Therefore

$4x + 2(x + 17) = 52$ ⇐ Remove the parentheses

$4x + 2x + 34 = 52$ ⇐ Combine like terms and subtract 34 from both sides

$6x = 18$ which implies that $x = \frac{18}{3} = 3$. Therefore the smaller number is 3 and the larger number is $3 + 17 = 20$.

Check: $4 \cdot 3 + 2 \cdot 20 = 12 + 40 = 52$. ✓

[57] To find a number which is 30 less than its opposite, let x be the number. The opposite of x is $-x$; and 30 less than the opposite of x is $-x - 30$. Therefore

$x = -x - 30$

$2x = -30$ or $x = -\frac{30}{2} = -15$.

Check: The opposite of -15 is 15. 30 less than the opposite is $15 - 30 = -15$. ✓

[61] Problem Statement: Find three consecutive integers whose sum is -66.

If x is the smallest integer then the next higher integer is $x + 1$, and the next after this is $x + 2$. Therefore x, $x + 1$, $x + 2$ are the 3 consecutive integers. Since their sum is -66,

$x + (x + 1) + (x + 2) = -66$ ⇐ Remove parentheses and combine like terms

$3x + 3 = -66$ ⇐ Subtract 3 from both sides

$3x = -66 - 3$ or $3x = -69$ ⇐ Divide both sides by 3

$x = -23$. Then $x + 1 = -23 + 1 = -22$, and $x + 2 = -23 + 2 = -21$.

Therefore the 3 consecutive integers are $-23, -22,$ and -21.

[65] Problem Statement: Find three consecutive even integers such that the largest is three times the smallest.

If x is the smallest of the integers, then $x + 2$ is the next even integer, and $x + 4$ is the next after that. Since the largest, $x + 4$, is three times the smallest, x, we have the equation

Exercises 2.4

$x + 4 = 3x$ ⇐ Subtract x from both sides

$4 = 2x$ ⇐ Divide both sides by 2

$2 = x$. Therefore the smallest of the even integers is 2, the next higher is $2 + 2 = 4$, and the largest is $2 + 4 = 6$.

69 Problem Statement: Find three consecutive odd integers such that three times the first minus the second plus four times the third is -4 .

If x is the smallest of the odd integers, then the next two higher odd integers are $x + 2$ and $x + 4$. Three times the first is $3x$, minus the second is $-(x + 2)$, and four times the third is $4(x + 4)$. Therefore $3x - (x + 2) + 4(x + 4) = -4$ which is equivalent to $3x - x - 2 + 4x + 16 = -4$ which reduces to $6x + 14 = -4$ or $6x = -18$ which implies that $x = -\frac{18}{6} = -3$. The next higher odd integer is $-3 + 2 = -1$, and the largest is $-3 + 4 = 1$. Therefore the 3 odd integers are -3, -1, and 1 .

73 To solve $3(2x - 1) - 4(8 - x) = 3x - (-x - 8) + 7$ remove the grouping symbols and simplify:

$6x - 3 - 32 + 4x = 3x + x + 8 + 7$

$10x - 35 = 4x + 15$ ⇐ Subtract $4x$ from both sides and add 35 to both sides

$6x = 50$ ⇐ Divide both sides by 6

$x = \frac{50}{6} = \frac{\cancel{2} \cdot 25}{\cancel{2} \cdot 3} = \frac{25}{3}$.

77 Problem statement: A quantity, its $\frac{2}{3}$, its $\frac{1}{2}$, and its $\frac{1}{7}$ added together becomes 33. Find the quantity.

If x is the quantity then

$x + \frac{2}{3}x + \frac{1}{2}x + \frac{1}{7}x = 33$ ⇐ Multiply both sides by $3 \cdot 2 \cdot 7 = 42$

$42\left(x + \frac{2}{3}x + \frac{1}{2}x + \frac{1}{7}x\right) = 42(33)$ ⇐ Use the distributive law to remove parentheses

$42x + 42 \cdot \frac{2}{3}x + 42 \cdot \frac{1}{2}x + 42 \cdot \frac{1}{7}x = 1386$ ⇐ Cancel the 3 into the 42, the 2 into the 42, and the 7 into the 42

$42x + 14 \cdot 2x + 21x + 6x = 1386$

$42x + 28x + 21x + 6x = 1386$

$97x = 1386$ or $x = \frac{1386}{97} \approx 14.29$.

Exercises 2.4

Review Problems

[85] $\frac{13}{100} = .13$

[89] $1\frac{2}{3} = 1.\overline{66}$

Exercises 2.5

[5] Problem Statement:
If an iceberg has a nearly uniform cross section, then its depth below the surface of the ocean is about 8 times its height above the surface. If the total height of an iceberg is 135 feet, how much is above and how much is below the surface?

Let x be the height above the surface. Then $8x$ is its depth below the surface. Since its total height is 135 feet, $8x + x = 135$ or $9x = 135$ which implies that $x = \frac{135}{9} = 15$.
Therefore 15 feet is above the surface and $8 \cdot 15 = 120$ feet is below the surface.

[9] Problem Statement:
The sum of the heights of Mt. Everest and Mt. Kilimanjaro is 48,460 feet. If Mt. Everest were 860 feet higher and Mt. Kilimanjaro were 4,320 feet lower, then Mt. Everest would be twice the height of Mt. Kilimanjaro. How high is each mountain?

Let x be the height of Mt. Everest. Then $48,460 - x$ is the height of Mt. Kilamanjaro.

The height of Mt. Everest plus an additional 860 feet is $x + 860$. The height of Mt. Kilimanjaro less 4,320 feet is $48,460 - x - 4,320 = 44,140 - x$. Therefore $x + 860 = 2(44,140 - x)$ or $x + 860 = 88,280 - 2x$ which simplifies to $3x = 87,420$ which implies that $x = \frac{87,420}{3} = 29,140$.

Therefore Mt. Everest is 29,140 feet high and Mt. Kilimanjaro is $48,460 - 29,140 = 19,320$ feet.

[17] Problem Statement:
Trifina paid $8.28 for 42 stamps. If she only bought 22¢ and 14¢ stamps, how many of each kind did she buy?

Let x be the number of 22¢ stamps. Since x plus the number of 14¢ stamps is 42, the number of 14¢ stamps is $42 - x$. The cost in pennies of the 22¢ stamps is 22 times the number purchased and the cost of the 14¢ stamps is 14 times the number purchased. Since the total cost in pennies is 828¢, x must satisfy the following equation:

$$22x + 14(42 - x) = 828 \quad \text{or} \quad 22x + 588 - 14x = 828 \quad \text{which simplifies to}$$
$$8x = 240 \quad \text{which implies that} \quad x = \frac{240}{8} = 30 \ .$$

Therefore the number of 22¢ stamps purchased is 30 and the number of 14¢ stamps purchased is $42 - 30 = 12$.

[21] Problem Statement:
Sean spent \$4.40 for some notebooks costing 80¢ each and pencils costing 10¢ each. If he bought 8 more pencils than notebooks, how many notebooks and pencils did he buy?

Let x be the number of notebooks. Then $x + 8$ is the number of pencils. Since the cost of the notebooks is $.80x$ and the cost of the pencils is $.10(x + 8)$, x must satisfy the following equation:

$$.80x + .10(x + 8) = 4.40 \quad \Leftarrow \text{ Multiply both sides by 10}$$
$$8x + 1 \cdot (x + 8) = 44 \quad \text{or} \quad 8x + x + 8 = 44 \quad \text{which reduces to}$$
$$9x = 36 \quad \text{which implies that} \quad x = \frac{36}{9} = 4 \ .$$

Therefore the number of notebooks is 4 and the number of pencils is $4 + 8 = 12$.

[25] Problem Statement:
A museum has three admission prices: \$1 for children, \$3.50 for adults, and \$1.50 for seniors. On Monday they sold 40 more senior tickets than adult tickets. If they sold 200 tickets for \$415, how many of each type of ticket were sold?

Let x be the number of adult tickets sold. Then $x + 40$ is the number of senior tickets.

Since the total number of tickets sold is 200, $x + (x + 40) + $ number of children's tickets $= 200$.

This means that the number of children's tickets is $200 - x - (x + 40)$ or $200 - x - x - 40$ which reduces to $160 - 2x$.

The x adult tickets sold are worth $3.50x$, the $x + 40$ senior tickets sold are worth $1.50(x + 8)$, and the $16 - 2x$ children's tickets sold are worth $1.00(160 - 2x)$. Since the total value is \$415,

$$3.50x + 1.50 \cdot (x + 40) + 1.00 \cdot (160 - 2x) = 415 \quad \Leftarrow \text{ Remove the parentheses}$$
$$3.5x + 1.5x + 60 + 160 - 2x = 415 \quad \Leftarrow \text{ Combine like terms and simplify}$$
$$3x = 195 \quad \text{or} \quad x = \frac{195}{3} = 65; \quad x + 40 = 105 \quad \text{and} \quad 160 - 2x = 160 - 130 = 30 \ .$$

Therefore 65 adult tickets, 105 senior tickets, and 30 children's tickets were sold.

[29] Problem Statement:
Five years from now Bruce will be three times as old as Chris was seven year ago. The sum of their current ages is 30. How old are they now?

Page 31

Exercises 2.5

Let x be Bruce's age now. Since $x+$ Chris's age is 30, Chris's age is $30-x$.
Five years from now, Bruce will be $x+5$ and 7 years ago, Chris was $30-x-7 = 23-x$.
Therefore $x+5 = 3(23-x)$ which reduces to $x+5 = 69 - 3x$ or $4x = 64$ which implies that $x = \frac{64}{4} = 16$. Therefore Bruce is now 16 years old and Chris is now $30-16 = 14$ years old.

33 Problem Statement:
Leah is 31 and Philip is 8. How long will it be before she is twice as old as he is?

Let x be the number of years for this to occur. By then, Leah will be $31+x$ and Philip will be $8+x$. Therefore, since Leah's age will be twice Philip's age, $31+x = 2(8+x)$ or $31+x = 16+2x$ which reduces to $15 = x$. So in 15 years, Leah will be twice Philip's age.

37 12.5% means $\frac{12.5}{100} = \frac{12.5}{100} \cdot \frac{10}{10} = \frac{125}{1000} = \frac{125 \cdot 1}{125 \cdot 8} = \frac{1}{8}$.

41 To change a number into a percent, move the decimal 2 places to the right, adding zeros if necessary. Therefore 62 expressed as a percent is 6200%.

45 $\frac{1}{5} = .2 = .20 = \frac{20}{100} = 20\%$.

49 Problem Statement: 30 is what percent of 125?

Use the fact that $\frac{\text{Percent}}{100} = \frac{\text{Part}}{\text{Whole}}$ where $30 = $ Part and $125 = $ Whole (Percent is unknown).
Let x be the percent. Then $\frac{x}{100} = \frac{30}{125}$ which is equivalent to $100 \cdot \frac{x}{100} = 100 \cdot \frac{30}{125}$ or $x = \frac{3000}{125} = 24$ percent.

53 Problem Statement: .23 is $\frac{1}{3}\%$ of what number?

Use the fact that $\frac{\text{Percent}}{100} \cdot \text{Whole} = \text{Part}$ where $.23 = $ Part and $\frac{1}{3} = $ Percent (Whole is unknown).
Let x be the number. Then $\frac{1/3}{100} x = .23$. Then multiply both sides by 100:
$100 \cdot \frac{1/3}{100} x = 100 \cdot (.23)$ or $\frac{1}{3}x = 23$. Finally multiply both sides by 3:
$3 \cdot \frac{1}{x} = 3 \cdot 23$ or $x = 69$.

Exercises 2.5

[57] The average of 74, 81, 72, 70, and 78 is $\dfrac{74+81+72+70+78}{5} = \dfrac{375}{5} = 75$.

[61] If her grades on the four tests are 71, 78, 86, and 82 and x is the grade on the final exam then her average is $\dfrac{71+78+86+82+x}{5}$ which must be at least 80. Therefore, at the very least,

$\dfrac{71+78+86+82+x}{5} = 80 \quad \Leftarrow$ Multiply both sides by 5

$71 + 78 + 86 + 82 + x = 400$ or $317 + x = 400$ which means that $x = 400 - 317 = 83$. Hence she needs a grade of 83.

[65] If an item priced at $30 was put on sale for 20% off then the sale price was $30 - .20(30) = 30 - 6 = \$24$.

[69] $2.5\% = \dfrac{2.5}{100} = .025$.

Therefore if we have 30 gallons of 2.5% salt solution then the amount of salt is 2.5% of 30 or $.025(30) = .75$ gallons of salt.

[73] The Schmitt's house sold for $90,850 and it increased in value by 15% over what they paid. Let x be the original cost. Then the increase in value is 15% of x or $.15x$. Therefore $x + .15x = 90,850$ or $1.15x = 90,850$. Divide both sides of the equation by 1.15:

$x = \dfrac{90,850}{1.15} = \$79,000$. Therefore they paid $79,000 for the house.

[77] An investment at 12.25% simple interest grows in one year to $2806.25. Let x be the amount invested. Then the increase is 12.25% of x or $.1225x$, and the total amount of money after one year is $x + .1225x = 1.00x + .1225x = 1.1225x$. Therefore $1.1225x = 2806.25$ and $x = \dfrac{2806.25}{1.1225} = \$2500$.

[81] Problem Statement: A carpenter has to cut a 20 foot board into 3 pieces. The second piece has to be 1 foot longer than the shortest piece and the third piece has to be 3 feet longer than twice the shortest piece. How long is each piece?

Let x be the length of the shortest piece. Then $x + 1$ is the length of the second piece.

Since the third piece is 3 feet longer than twice the shortest piece, the length of the 3rd piece is

Exercises 2.5

$3 + 2x$. The sum of the 3 lengths must be 20 feet. Therefore
$x + (x+1) + (2x+3) = 20$ which simplifies to $4x + 4 = 20$ or $4x = 16$ which means that $x = \frac{16}{4} = 4$. So the shortest piece is 4 feet, the 2nd piece is 5 feet, and the longest piece is $2 \cdot 4 + 3 = 11$ feet.

[85] Problem Statement: Thomas Jefferson was born 11 years after George Washington. In 1762 Washington's age was 3 years more than 3 times Jefferson's age in 1752. How old was each man in 1755?

Let x be Thomas Jefferson's age in 1752.

In 1762 George Washington was $3 + 3x$.

In 1762 Thomas Jefferson was $x + 10$. ⇐ $1762 - 1752 = 10$

Since George Washington was 11 years older than Thomas Jefferson, in 1762 we see that

$3 + 3x = x + 10 + 11$ which simplifies to

$2x = 18$ or $x = 9$. So in 1752 Thomas Jefferson was 9 and George Washington was 20.

Ans: In 1755, Jefferson was $9 + 3 = 12$ and Washington was $20 + 3 = 23$.

Review Problems

[93] $5x + 4 = 2x + 11$
$5x + 4 - 4 - 2x = 2x + 11 - 4 - 2x$
$3x = 7$ or $x = \frac{7}{3}$.

[97] $\frac{7}{11} = .63\overline{63}$

Exercises 2.6

[1] If Teresa drove for 6 hours at an average speed of 38 mph then
$D = RT = 38 \cdot 6 = 228$ miles.

[5] If Ed hiked 100 km and the trip took 16 hours, his average rate was
$R = \frac{D}{T} = \frac{100 \text{ km}}{16 \text{ hrs}} = \frac{\cancel{4} \cdot 25}{\cancel{4} \cdot 4}$ km/hr $= \frac{25}{4}$ km/hr or $6\frac{1}{4}$ km/hr.

Exercises 2.6

[9] To find the length of a rectangle if its width is 6 ft and its area is 51 ft^2, use the fact that $A = L \times W$ where A is the area, L is the length, and W is the width. Then $6L = 51$ which implies that $L = \frac{51}{6} = 8\frac{1}{2}$ ft .

[13] The perimeter of a triangle is the sum of the lengths of the 3 sides. If two of the sides are 18 inches and 23 inches and we let x be the length of the 3rd side then $18 + 23 + x = 67$ or $41 + x = 67$ which implies that $x = 67 - 41 = 26$ inches.

[17] If $V = LWH$ and $V = 60$, $L = 3$, and $H = 4$ then
$60 = 3 \cdot W \cdot 4$ or $60 = 12W$ which implies that $W = \frac{60}{12} = 5$.

[21] If $y = -\frac{2}{3}x + 5$ and $x = -6$ then $y = -\frac{2}{3}(-6) + 5 = \frac{-2}{3} \cdot \frac{-6}{1} + 5 = \frac{12}{3} + 5 = 4 + 5 = 9$.

[25] If $A = P + Prt$ and $A = 248$, $r = 8\%$, and $t = 3$ then

$248 = P + P(.08)(3)$ ⇐ Note that $8\% = \frac{8}{100} = .08$

$248 = P + .24P$ ⇐ Think of P as 1P

$248 = (1 + .24)P$ or $248 = 1.24P$ which implies that $P = \frac{248}{1.24} = 200$.

[29] If $S = 2ab + 2ac + 2bc$ and $S = 88$, $a = 6$, and $b = 4$ then

$88 = 2(6)(4) + 2(6)c + 2(4)c$ which simplifies to

$88 = 48 + 12c + 8c$ or $88 = 48 + 20c$, and add -48 to both sides,

$40 = 20c$ which implies that $c = \frac{40}{20} = 2$.

[33] To solve $A = \pi r^2 h$ for h, divide both sides of the equation by πr^2 so that h is isolated:
$\frac{A}{\pi r^2} = \frac{\pi r^2 h}{\pi r^2}$ or $h = \frac{A}{\pi r^2}$.

[37] To solve $3x - 4y = 12$ for x, first isolate the $3x$ term by adding $4y$ to both sides:

$3x - 4y + 4y = 12 + 4y$

$3x = 4y + 12$ ⇐ Divide both sides by 3 to isolate y

$\frac{3x}{3} = \frac{4y + 12}{3}$ ⇐ Don't write $\frac{4y}{3} + 12$ or $4y + 4$

$x = \frac{4y + 12}{3}$.

Exercises 2.6

41 To solve $A = P + Prt$ for P, use the distributive law to write $P + Prt$ as $P(1 + rt)$:

$A = P(1 + rt) \quad \Leftarrow$ Divide both sides by $1 + rt$ to isolate the P

$\dfrac{A}{1+rt} = \dfrac{P(1+rt)}{1+rt} \quad \Leftarrow$ Cancel the $1 + rt$ in the numerator and denominator

$\dfrac{A}{1+rt} = P \quad$ or $\quad P = \dfrac{A}{1+rt}$.

45 To solve $C = \dfrac{5}{9}(F - 32)$ for F, simplify the equation by multiplying both sides by $\dfrac{9}{5}$:

$\dfrac{9}{5}C = \dfrac{9}{5} \cdot \dfrac{5}{9}(F - 32) \quad \Leftarrow \dfrac{9}{5} \cdot \dfrac{5}{9} = 1$

$\dfrac{9}{5}C = F - 32 \quad \Leftarrow$ Add 32 to both sides to isolate the F

$\dfrac{9}{5}C + 32 = F \quad$ or $\quad F = \dfrac{9}{5}C + 32$.

49 To solve $L = a + (n - 1)d$ for d, subtract a from both sides:

$L - a = (n - 1)d \quad \Leftarrow$ Divide both sides by $n - 1$ to isolate the d

$\dfrac{L-a}{n-1} = \dfrac{(n-1)d}{n-1} \quad \Leftarrow$ Cancel the $n - 1$

$\dfrac{L-a}{n-1} = d \quad$ or $\quad d = \dfrac{L-a}{n-1}$

53 To solve $y = \dfrac{3x-1}{4}$ for x, simplify the equation by multiplying both sides by 4:

$4y = 4 \cdot \dfrac{3x-1}{4} \quad \Leftarrow$ Cancel the 4's on the right side

$4y = 3x - 1 \quad \Leftarrow$ Add 1 to both sides

$4y + 1 = 3x \quad \Leftarrow$ Divide both sides by 3

$\dfrac{4y+1}{3} = \dfrac{3x}{3} \quad$ or $\quad x = \dfrac{4y+1}{3}$.

57 $.33 = \dfrac{33}{100}$

61 To see how to express $.21\overline{21}$ as a ratio of 2 integers, let $x = .21\overline{21}$.
Then $100x = 21.21\overline{21}$. Now subtract x from $100x$:
$100x - x = 21.21\overline{21} - .21\overline{21} = 21$. But $100x - x = 99x$. Therefore
$99x = 21$ which implies that $x = \dfrac{21}{99} = \dfrac{3 \cdot 7}{3 \cdot 33} = \dfrac{7}{33}$. Ans: $.21\overline{21} = \dfrac{7}{33}$.

Exercises 2.6

65 Let $x = .24\overline{6246}$. Then $1000x = 246.24\overline{6246}$. Subtract x from $1000x$:

$1000x - x = 246.24\overline{6246} - .24\overline{6246} = 246$. But $1000x - x = 999x$. Therefore

$999x = 246$ which implies that $x = \frac{246}{999} = \frac{3 \cdot 82}{3 \cdot 333} = \frac{82}{333}$. Ans: $.24\overline{6246} = \frac{82}{333}$.

69 Let $x = 1.247\overline{47}$. Then $10x = 12.47\overline{47}$ and $1,000x = 1247.47\overline{47}$. Subtract $10x$ from $1000x$:

$990x = 1247.47\overline{47} - 12.47\overline{47}$ or $990x = 1235$. This means that

$x = \frac{1235}{990} = \frac{5(247)}{5(198)} = \frac{247}{198}$. Ans: $1.247\overline{47} = \frac{247}{198}$.

73 Let $x = .249\overline{9}$. Then $1000x = 249.9\overline{9}$ and $100x = 24.9\overline{9}$.

$1000x - 100x = 249.9\overline{9} - 24.9\overline{9} = 225$

$900x = 225$ or $x = \frac{225}{900} = \frac{225}{225 \cdot 4} = \frac{1}{4}$. Ans: $.249\overline{9} = \frac{1}{4}$.

77 If $S = P(1+i)^n$ and $P = 100$, $i = 10\% = \frac{10}{100} = .1$, and $n = 2$ then

$S = 100(1 + .10)^2 = 100(1.1)^2 = 100(1.21) = 121$.

81 To solve $W = k(L+4) + T$ for L first remove the parentheses:

$W = kL + 4k + T$ ⇐ Add $-4k - T$ to both sides to isolate the kL term

$W - 4k - T = kL$ ⇐ Divide both sides by k to isolate the L

$\frac{W - 4k - T}{k} = \frac{kL}{k}$ or $L = \frac{W - 4k - T}{k}$.

85 To solve $v = \frac{S-s}{t}$ for t, multiply both sides by t:

$vt = \frac{S-s}{t} \cdot t$ ⇐ Cancel the t's on the right side

$vt = S - s$ ⇐ Divide both sides by v

$t = \frac{S-s}{v}$.

Review Problems

93 If x is the first of three consecutive integers, the next two are $x+1$ and $x+2$.

Exercises 2.7

[5] Problem Statement: Joe and Sandra live 21 miles apart. At 8 AM Sandra started jogging toward Joe's house at a rate of 5mph. At 9 AM Joe started walking toward Sandra's house at a rate of 3 mph. When will they meet?

Let t be the time Sandra traveled. Then $t-1$ is the time Joe traveled. Since D=RT,

Sandra traveled a distance $5t$ and Joe traveled a distance $3(t-1)$. Since the total must be 21,

$5t + 3(t-1) = 21$ which simplifies to $5t + 3t - 3 = 21$ or $8t = 24$ which means that

$t = \frac{24}{8} = 3$. Since Sandra started at 8 AM, and $3 + 8 = 11$, they meet at time 11 AM.

[9] Problem Statement: Bryan and Randy start a trip from the same point at the same time and travel in the same direction. Randy cycles at 4/5 Bryan's average speed. After 2 hours and 15 minutes Bryan is 9 km ahead of Randy. How fast does each cycle?

Let x be Bryan's speed. Then $\frac{4}{5}x$ is Randy's speed. 2 hours and 15 minutes later, Bryan travels a distance $\left(2\frac{1}{4}\right)x$ and Randy travels a distance $\left(2\frac{1}{4}\right)\left(\frac{4}{5}\right)x$. Since Bryan's distance is 9 km more than Randy's distance, we have the equation:

$\left(2\frac{1}{4}\right)x = \left(2\frac{1}{4}\right)\left(\frac{4}{5}\right)x + 9$ or $\frac{9}{4}x = \frac{9}{4} \cdot \frac{4}{5}x + 9$ which reduces to $\frac{9}{4}x = \frac{9}{5}x + 9$. After multiplying both sides by 20 we get $20 \cdot \frac{9}{4}x = 20 \cdot \frac{9}{5}x + 20 \cdot 9$ which after canceling the 4 into the 20 and canceling the 5 into the 20 simplifies to

$45x = 36x + 180$ or $9x = 180$ which means that $x = \frac{180}{9} = 20$.

Therefore Bryan's speed is $20 \frac{km}{hr}$ and Randy's speed is $\frac{4}{5} \cdot 20 = 16 \frac{km}{hr}$.

[13] Problem Statement: Rosa took 10 hours to drive 396 miles. During part of the trip she averaged 30 mph and during the remainder of the trip she averaged 46 mph. How far did she travel at each rate?

Let x be the time going 30 mph and then $10 - x$ is the time going 46 mph. Since the total distance is 396 miles and D = RT,

$30x + 46(10-x) = 396$ which simplifies to $30x + 460 - 46x = 396$ or $-16x = -64$ which means that $x = \frac{-64}{-16} = 4$ and $10 - x = 6$.

Therefore at 30 mph she traveled $30 \cdot 4 = 120$ miles and at 46 mph she traveled $46 \cdot 6 = 276$ miles.

[17] Problem Statement: Felba can paddle her canoe at a rate of 5 mph in still water. She paddles downstream for 3 hours. On the return trip she paddles upstream for 5 hours and is still 6 miles from her starting point. What is the speed of the stream? How far downstream did she travel?

Let x be the speed of the stream. Then her speed downstream is $5 + x$ and her speed upstream is $5 - x$. Since she spends 3 hours traveling downstream, the distance she goes downstream is $3(5 + x)$. In 5 hours going upstream, she travels a distance $5(5 - x)$.

Since $5(5 - x)$ is 6 miles less than $3(5 + x)$ we have the following equation:

$3(5 + x) = 5(5 - x) + 6$ $\Leftarrow$ Note that we add 6 to the smaller quantity to get equal quantities

This simplifies to $15 + 3x = 25 - 5x + 6$ or $8x = 16$ which means $x = 2$.

Therefore the speed of the stream is 2 mph and Felba travels downstream a distance equal to $3(5 + 2) = 3 \cdot 7 = 21$ miles.

[21] If the length and width are consecutive even integers then we can denote the length and width by $x + 2$ and x. Since the perimeter is 44 inches and the perimeter equals twice the length plus twice the width we have the equation:

$2(x + 2) + 2x = 44$ or $2x + 4 + 2x = 44$ which reduces to $4x = 40$ which means that $x = \frac{40}{4} = 10$. Therefore the width is 10 inches and the length is 12 inches. The area is length $\times$ width or $10 \cdot 12 = 120$ square inches.

[25] The area of a trapezoid is given by $A = \frac{1}{2}h(b_1 + b_2)$ where h is the height and b_1 and b_2 are the two bases. We know one base is 16, call it b_1, and since we don't know b_2, call it x.

Since $h = 6$ and $A = 108$ we have the equation:

$108 = \frac{1}{2}(6)(16 + x)$ or $108 = 3(16 + x)$ which reduces to $108 = 48 + 3x$ which means that $60 = 3x$ or $x = \frac{60}{3} = 20$.

Exercises 2.7

29 Let x be the length of the base. Since the 2 equal sides are 4 cm longer than the base, their length is $x+4$. Since the perimeter, the sum of the 3 lengths, is 20 cm, we have the equation: $x+2(x+4)=20$ or $x+2x+8=20$ which simplifies to $3x=12$ or $x=4$.
Therefore the base is 4 cm.

37 Problem Statement: Marlene invested $8000, part in bonds at 6% and the rest in a mutual fund at 9%. The annual interest earned at 6% was twice the annual interest earned at 9%. How much did she invest at each rate?

(Note the symbol $\Rightarrow$ means implies)

Let x be the amount invested at 9%. Then $8000-x$ is the amount invested at 6%.

The interest earned on the amount x is $.09x$ and the interest earned on the amount $8000-x$ is $.06(8000-x)$. Therefore

$.06(8000-x) = 2(.09x) \Rightarrow 480 - .06x = .18x \Rightarrow 480 = .06x + .18x \Rightarrow$
$480 = (.06+.18)x \Rightarrow 480 = .24x \Rightarrow \frac{480}{.24} = \frac{.24x}{.24} \Rightarrow x = \frac{480}{.24} = 2000$.

Therefore she invested $2000 at 9% and $8000-2000 = \$6000$ at 6%.

45 Problem Statement: A shop mixed cashews worth $8 per kg with pecans worth $9 per kg to produce a 10 kg mixture worth $84. How many kilograms of each are in the mixture?

Let x be the number of kilograms of cashews and $10-x$ the number of kilograms of pecans. Then the x kg of cashews are worth $8x$ and the $10-x$ kilograms of pecans are worth $9(10-x)$. Therefore the total mixture is worth $8x + 9(10-x)$ and we have the equation

$8x + 9(10-x) = 84 \Rightarrow 8x + 90 - 9x = 84 \Rightarrow -x = -6 \Rightarrow x = 6$.

Hence the mixture contains 6 kg of cashews and 4 kg of pecans.

53 Problem Statement: A certain metal is 25% copper. How many kg of this metal must be melted with 80 kg of 40% copper to obtain a metal which is 30% copper?

Let x be the amount of metal. Then $.25x$ is the amount of copper in the x kilograms. In 80 kg of 40% copper there is $.40(80) = 32$ kg of copper. Therefore in the mixture, there is $.25x$ kg $+ 32$ kg of copper. Since the mixture which totals $x+80$ must be 30% copper, the amount of copper in the mixture should be $.30(x+80)$. Therefore we have the equation

$.25x + 32 = .30(x+80) \Rightarrow .25x + 32 = .30x + 24 \Rightarrow 8 = .05x \Rightarrow x = \frac{8}{.05} = 160$.

Hence there is 160 kg of this metal.

Exercises 2.7

57. Problem Statement: How much distilled water has to be added to 25 liters of a 20% acid solution to dilute it to a 12% solution?

Let x be the amount of water. Then the total mixture is $x + 25$. Since no more acid is added, the mixture contains $(.20)(25) = 5$ liters of acid, the amount of acid in the original 25 liters. Since we want a 12% solution, the amount of acid in the mixture should be $.12(x+25)$.

Therefore we have the equation

$.12(x+25) = 5 \Rightarrow .12x + 3 = 5 \Rightarrow .12x = 2 \Rightarrow x = \frac{2}{.12} = 16.67$ liters.

61. Problem Statement (Example 9): A chemist has 60 cc of a 35% acid solution. How much distilled water should be added to dilute it to a 20% acid solution? Solve by calculating the amount of water before and after adding the distilled water.

Since the mixture is 35% acid, it is 65% water. Then the amount of water at the start is $.65(60) = 39$ cc. The amount of acid is $.35(60) = 21$ cc. Note that $39 + 21 = 60$.

Let x be the amount of water after we add the distilled water. Then the mixture contains a total of $x + 21$ cc (the acid plus the water). Since the mixture is 20% acid and the mixture has 21 cc of acid, we have the equation

$.20(x+21) = 21 \Rightarrow .20x + 4.2 = 21 \Rightarrow .20x = 16.8 \Rightarrow x = \frac{16.8}{.20} = 84$.

Therefore there is a total of 84 cc of water in the mixture which means that $84 - 39 = 45$ cc of distilled water was added to dilute the mixture.

65. Problem Statement: In the Rhind papyrus, the area of a circle is taken to be the square of 8/9 of the diameter. Use the formula $A = \pi r^2$ to determine the ancient Egyptian value of π, correct to two decimal places. How does this compare with our decimal value of π?

$A = \pi r^2 = \left(\frac{8}{9}d\right)^2 = \left(\frac{8}{9} \cdot 2 \cdot r\right)^2 = \left(\frac{16}{9}\right)^2 r^2$. Therefore $\pi r^2 = \left(\frac{16}{9}\right)^2 r^2 \Rightarrow \pi = \left(\frac{16}{9}\right)^2 \Rightarrow \pi = 3.16$.

Review Problems

73. $-5 \leq -1$ can be rewritten as $-1 \geq -5$.

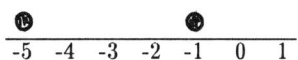

Exercises 2.7

77 If x is the number then "A number is greater than -4" is expressed as $x > -4$.

Exercises 2.8

1 $x \leq 2$ translates into English as "x is less than or equal to 2".

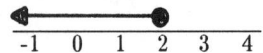

5 $-2 < n < 2$ translates into English as "n is greater than -2 and n is less than 2".

9 $-1 < t \leq \frac{1}{2}$ translates into English as "t is greater than -1 and t is less than or equal to $\frac{1}{2}$".

13 $-2 \geq -6$ and $-3 + (-2) \geq -3 + (-6)$.

17 $-2 \geq -6$ and $\frac{-2}{-3} \leq \frac{-6}{-3}$. Note that when we multiply or divide both sides of an inequality by a negative number, the sense of the inequality is reversed.

21 To solve $x + 4 < 6$ subtract 4 from both sides:

$x + 4 - 4 < 6 - 4$ which implies that

$x < 2$.

25 To solve $-3x < 12$ divide both sides by -3, remembering to reverse the sense of the inequality:

$\frac{-3x}{-3} > \frac{12}{-3} \Rightarrow x > -4$.

Exercises 2.8

[29] To solve $6 \geq 2x > -8$, divide each of the 3 quantities by 2:

$\frac{6}{2} \geq \frac{2x}{2} > \frac{-8}{2}$ which implies that

$3 \geq x > -4$ or equivalently

$-4 < x \leq 3$.

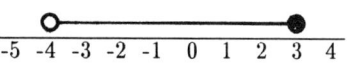

[33] To solve $3 - 2y \leq -17$, subtract 3 from both sides:

$3 - 2y - 3 \leq -17 - 3$

$-2y \leq -20$ ⇐ Divide both sides by -2 reversing the sense of the inequality

$\frac{-2y}{-2} \geq \frac{-20}{-2}$ or $y \geq 10$.

[37] To solve $1 + \frac{m}{5} - 6 < -2$, combine the 1 and -6:

$-5 + \frac{m}{5} < -2$ ⇐ Add 5 to both sides

$5 - 5 + \frac{m}{5} < -2 + 5$

$\frac{m}{5} < 3$ ⇐ Multiply both sides by 5

$m < 15$

[41] To solve $2x - (10 - x) \geq 8x + 1$ remove the parentheses and simplify:

$2x - 10 + x \geq 8x + 1$

$3x - 10 \geq 8x + 1$ ⇐ Subtract $3x$ from both sides

$-10 \geq 5x + 1$ ⇐ Subtract 1 from both sides

$-11 \geq 5x$ ⇐ Divide both sides by 5

$\frac{-11}{5} \geq \frac{5x}{5}$ or $-\frac{11}{5} \geq x$ which is equivalent to $x \leq -\frac{11}{5}$.

[45] To solve $-2 \leq 4 - k \leq 2$ subtract 4 from each of the 3 quantities:

$-2 - 4 \leq 4 - k - 4 \leq 2 - 4$

$-6 \leq -k \leq -2$ ⇐ Multiply each quantity by -1 and reverse the sense of

$6 \geq k \geq 2$ the inequalities.

which is equivalent to $2 \leq k \leq 6$.

Exercises 2.8

49 To solve $-7 \leq \frac{t}{6}+2 \leq 9$ subtract 2 from each of the quantities:

$$-7-2 \leq \frac{t}{6}+2-2 \leq 9-2$$

$$-9 \leq \frac{t}{6} \leq 7 \qquad \Leftarrow \text{ Multiply each quantity by 6}$$

$$-9 \cdot 6 \leq \frac{t}{6} \cdot 6 \leq 7 \cdot 6 \quad \text{or} \quad -54 \leq t \leq 42 \, .$$

53 If his time in the race (t) was no more than 3 hours then $t \leq 3$.

57 If the speed of the car (r) is between is 50 and 60 mph then $50 < r < 60$.

61 Problem Statement: The sum of three consecutive even integers is greater than 69. Find the smallest possible values for the integers.

Let x, $x+2$, and $x+4$ be the integers. Then $x+(x+2)+(x+4) > 69$ which simplifies to $3x+6 > 69$ or $3x > 63$ which means that $x > 21$. Therefore the smallest possible values are 22, 24, and 26 .

65 Problem Statement: The average of four consecutive odd integers is not greater than or equal to -9 . Find the largest possible values for the integers.

Let x, $x+2$, , $x+4$, and $x+6$ be the four integers. The average is the sum of the 4 numbers divided by 4 or $\frac{x+x+2+x+4+x+6}{4}$. Note that "not greater than or equal to -9" means "less than -9" .

Therefore $\frac{x+x+2+x+4+x+6}{4} < -9$ or $\frac{4x+12}{4} < -9$ and after multiplying both sides by 4 we get: $4x+12 < -36$ or $4x < -48$ which implies that $x < -12$.
If $x < -12$ then the largest possible odd integer value for x is -13 . Therefore the 4 integers are -13, -11, -9, and -7 .

69 Problem Statement: Liquid penicillin has to be kept between 5°C and 15°C, inclusive. Find the range on the Fahrenheit scale; $C = \frac{5}{9}(F-32)$.

Exercises 2.8

Therefore $5 \leq \frac{5}{9}(F-32) \leq 15$. To simplify, multiply each of the 3 quantities by $\frac{9}{5}$:

$$\frac{9}{5} \cdot 5 \leq \frac{9}{5} \cdot \frac{5}{9}(F-32) \leq \frac{9}{5} \cdot 15$$

$9 \leq F - 32 \leq 27 \quad \Leftarrow$ Add 32 to each of the quantities

$41 \leq F \leq 59$. Hence the range on the Fahrenheit scale is between 41°F and 59°F, inclusive.

[73] Problem Statement: Javier was given $2000 for a High School graduation present. He wants to pay for a car and the 6.5% sales tax out of the $2000. What is the most amount of money (to the nearest dollar) that he can spend for the car?

Let x be the amount he spends for the car. Then the total cost, including the tax, is

$x + .065x \quad \Leftarrow$ Note that $6.5\% = \frac{6.5}{100} = .065$

$x + .065x$ can not exceed 2000. Therefore $x + .065x \leq 2000$ or $1.065x \leq 2000$.

Then $x \leq \frac{2000}{1.065}$ or $x \leq 1877.90$. So the most he can spend on the car is $1878.

[77] Problem Statement: A house and a lot together cost at least $220,000. The house costs $10,000 more than 6 times the cost of the lot. What is the least amount that the lot cost?

Let x be the cost of the lot. Then $6x + 10,000$ is the cost of the house.

Since the total cost is at least 220,000, we have $x + 6x + 10,000 \geq 220,000$ which reduces to $7x \geq 210,000$ or $x \geq 30,000$. Therefore the lot must cost at least $30,000.

[81] Problem Statement: The length of a rectangle is one more than twice the width. The perimeter must be more than 55 m. Find the minimum dimensions of the rectangle if they are integers.

Let x be the width. Then $2x + 1$ is the length. Since the perimeter is twice the length plus twice the width, $2x + 2(2x + 1) > 55$ which simplifies to $6x + 2 > 55 \Rightarrow 6x > 53$. Hence $x > \frac{53}{6} = 8.8$. If x must be an integer, then the smallest x can be is 9 m and $2x + 1 = 19$.

Therefore the minimum dimensions are 9 m × 19 m.

[85] Problem Statement: Three-fourths of John's age three years ago is at least three-fifths of his age one year from now. At least how old is he now?

Let x be John's age now. His age three years ago is $x - 3$ and his age one year from now is $x + 1$. Then three-fourths of John's age three years ago is $\frac{3}{4}(x - 3)$ and three-fifths of his age one year

Page 45

from now is $\frac{3}{5}(x+1)$. Therefore $\frac{3}{4}(x-3) \geq \frac{3}{5}(x+1)$. To simplify the inequality, multiply both sides by $4 \times 5 = 20$ and cancel the 4 into the 20 and the 5 into the 20:

$20 \cdot \frac{3}{4}(x-3) \geq 20 \cdot \frac{3}{5}(x+1) \Rightarrow 5 \cdot 3(x-3) \geq 4 \cdot 3(x+1) \Rightarrow$

$15x - 45 \geq 12x + 12 \Rightarrow 3x \geq 57 \Rightarrow x \geq 19$. So John is at least 19 years old.

[89] To solve $3 + \frac{x+1}{2} > 1$ subtract 3 from both sides which gives

$\frac{x+1}{2} > -2$ $\Leftarrow$ Multiply both sides by 2 to eliminate the fraction

$x + 1 > -4$ $\Leftarrow$ Subtract 1 from both sides

$x > -5$.

[93] To solve $x(x-4) > x^2 + x + 1$ remove the grouping symbols:

$x^2 - 4x > x^2 + x + 1$ $\Leftarrow$ Subtract x^2 from both sides

$-4x > x + 1$ $\Leftarrow$ Subtract x from both sides

$-5x > 1$ $\Leftarrow$ Divide both sides by -5 and reverse the sense of the

$x < -\frac{1}{5}$.   inequality

[97] $(1.4)^2 - 4.1x \leq -3.1 - 2^3$ simplifies to $1.96 - 4.1x \leq -3.1 - 8$ or $1.96 - 4.1x \leq -11.1$

Then subtract 1.96 from both sides which gives $-4.1x \leq -13.06$ which implies that

$x \geq \frac{-13.06}{-4.1}$ or $x \geq 3.19$.

Chapter Review Exercises

1. $\dfrac{20+33+41}{3} = \dfrac{94}{3} \approx 31.33$.

5. Let $x = .\overline{77}$. Then $10x = 7.\overline{77}$ and $10x - x = 7.\overline{77} - .\overline{77} = 7$. Therefore $9x = 7$ which implies that $x = \dfrac{7}{9}$.

9. The graph of $x > 4$ is

    ```
    ─────○──────▶
    0 1 2 3 4 5 6 7 8 9
    ```

13. $2(3x+y)+4(x-6y) = 6x+2y+4x-24y = (6x+4x)+(2y-24y) = 10x-22y$.

17. $3x - 2y = 12$
    $3x - 12 = 2y$
    $\dfrac{3x-12}{2} = y$ or $y = \dfrac{3x-12}{2}$.

21. $3m + 2 \geq -10$
    $3m \geq -12$
    $m \geq -4$.

25. $-8 < 1 - 2x \leq 10$
    $-9 < -2x \leq 9$
    $\dfrac{-9}{-2} > \dfrac{-2x}{-2} \geq \dfrac{9}{-2}$
    $\dfrac{9}{2} > x \geq -\dfrac{9}{2}$ or $-\dfrac{9}{2} \leq x < \dfrac{9}{2}$.

29. $3(4x-1)+5 \leq -2(1-x)$
    $12x - 3 + 5 \leq -2 + 2x$
    $12x + 2 \leq 2x - 2$
    $10x \leq -4$ or $x \leq \dfrac{-4}{10}$ or $x \leq -\dfrac{2}{5}$.

33. Let x be the distance. Then $\dfrac{3}{5}x + 30 + \dfrac{1}{10}x = x$. To simplify this equation, multiply both sides by 10 which gives $10 \cdot \left(\dfrac{3}{5}x + 30 + \dfrac{1}{10}x\right) = 10x$ or $6x + 300 + x = 10x$ which reduces to $300 = 3x$ or $x = 100$ miles .

37. Let x be the percent. Then $\dfrac{x}{100}(350) = 42$. Multiply both sides by 100 : $350x = 4200 \Rightarrow x = \dfrac{4200}{350} = 12$. Therefore the percentage is 12% .

41. Let x be the shortest side. Then $3x$ is the 2nd side and $3x+2$ is the 3rd side and $x+3x+3x+2=30$ which reduces to $7x+2=30$ or $7x=28$. Hence $x=4$. Therefore the shortest side is 4 cm, the 2nd side is 12 cm, and the 3rd side is 14 cm.

Chapter Test

1. $1\frac{3}{4}\% = 1.75\% = .0175 = \frac{1.75}{100} = \frac{175}{10,000} = \frac{7\cdot 25}{400\cdot 25} = \frac{7}{400}$.

5. $A = \frac{1}{2}h(B+b) \Rightarrow 60 = \frac{1}{2}\cdot 6(B+8) \Rightarrow 60 = 3B+24 \Rightarrow 36 = 3B \Rightarrow B = 12$.

9. $4-2x \geq 22 \Rightarrow -2x \geq 18 \Rightarrow x \leq \frac{18}{-2} \Rightarrow x \leq -9$.

13. $10k-5 = 8-3k$
    $13k = 13$
    $k = 1$

17. $10-x > 4(x-6) - 3(4x+1)$
    $10-x > 4x - 24 - 12x - 3$
    $7x > -37$
    $x > -\frac{37}{7}$.

21. Let x be Matt's age. Then Cindy's age is $x+5$. Therefore $x+10 = 2(x+5-10)$ which simplifies to $x+10 = 2x-10 \Rightarrow 20 = x$. Therefore Matt is 20 and Cindy is 25.

25. Let x be the number of liters of 30% solution. In the 5 liters of 20% solution there is .20(5) or 1 liter of acid, and in the x liters there is .30x liters of acid. Then the mixture of $x+5$ liters has a total of .30$x+1$ liters of acid. Since the mixture must be 28% acid, the following equation must be true:

$.30x + 1 = .28(x+5) \Rightarrow .30x + 1 = .28x + 1.4 \Rightarrow .02x = .4 \Rightarrow$
$x = \frac{.4}{.02} = 20$ liters.

Chapter 3: Polynomials

Exercises 3.1

[1] $5 \cdot 5 \cdot 5 = 5^3$

[5] $0^{90} = 0$

[9] $x^3 \cdot x^5 = x^{3+5} = x^8$. Therefore ? = 8.

[13] $\dfrac{5^{10}}{5^{10}} = 5^{10-10} = 5^0 = 1$.

[17] $\left(\dfrac{x}{y}\right)^4 = \dfrac{x^4}{y^4}$. Therefore ? = 4.

[21] $3^2 \cdot 3^8 = 3^{16}$ is false; $3^2 \cdot 3^8 = 3^{2+8} = 3^{10}$.

[25] $-3^4 = -(3^4) = -81$. Therefore the statement is true.

[29] $\dfrac{7^5}{7^3} = 1^2$ is false; $\dfrac{7^5}{7^3} = 7^{5-3} = 7^2 = 49$.

[33] $(5+4)^2 = 9^2 = 81$.

[37] $\dfrac{4^{20}}{4^{17}} = 4^{20-17} = 4^3 = 64$.

[41] $3^4 \cdot 3^7 = 3^{11} = 177{,}147$.

[45] $3^8(-2)^8 = [3 \cdot (-2)]^8 = (-6)^8 = 1{,}679{,}616$

[49] $-(3y^n)^4 = -\left[3^4(y^n)^4\right] = -81y^{4n}$.

[53] $\dfrac{k^{20}}{k^{20}} = 1$

[57] $\dfrac{(b^5)^4}{b} = \dfrac{b^{20}}{b^1} = b^{20-1} = b^{19}$.

[61] $7^2 \cdot 7^{2n+10} = 7^{2+2n+10} = 7^{2n+12}$.

[65] $\dfrac{9^{6k-4}}{9^{2k-7}} = 9^{6k-4-(2k-7)} = 9^{6k-4-2k+7} = 9^{4k+3}$.

[69] $\dfrac{-x^2}{(-x)^6} = \dfrac{-x^2}{(-1)^6 x^6} = -\dfrac{x^2}{1 \cdot x^6} = -\dfrac{1}{x^{6-2}} = -\dfrac{1}{x^4}$.

Review Problems

[73] The coefficient of $12x^2$ is 12.

[77] $7xy^2$ and $2x^2y$ are unlike terms.

[81] $2x^2 - 3x + 3x^2 = 5x^2 - 3x$.

Exercises 3.2

1 The degree of $4z^2$ is 2 and its coefficient is 4.

5 The degree of $-x^2y^3$ is the sum of the powers of x and y or $2+3=5$. The coefficient is -1,

9 $(3x^2)(2x^5) = (3\cdot 2)(x^2 x^5) = 6x^7$.

13 $(2a^6)^3 = 2^3 \cdot (a^6)^3 = 8a^{18}$.

17 $\dfrac{8x^9}{2x^4} = 4x^{9-4} = 4x^5$.

21 $\dfrac{(2x^3)^3}{-6x^6} = \dfrac{2^3(x^3)^3}{-6x^6} = \dfrac{8x^9}{-6x^6} = -\dfrac{4\cdot 2}{3\cdot 2}x^{9-6} = -\dfrac{4}{3}x^3$.

25 $3k^2 + 5k^2 + k = 8k^2 + k$.

29 $-2x^4 + 8x^4 \neq 6x^8;\ -2x^4 + 8x^4 = 6x^4$.

33 $8x - y \neq 7xy;\ 8x - y$ can not be simplified.

37 $5a^3 + 3a^3 \neq 8a^9;\ 5a^3 + 3a^3 = 8a^3$.

41 $6a^3 - 2a^3 = (6-2)a^3 = 4a^3$.

45 $3rt + 6tr = 3rt + 6rt = 9rt$.

49 $(3x^4 y^3)^2 = 3^2(x^4)^2(y^3)^2 = 9x^8 y^6$.

53 $3xy^2 + 7x^2y - 4xy^2 = (3xy^2 - 4xy^2) + 7x^2y = -xy^2 + 7x^2y$.

57 $(-3x)(-2x) + 7x^2 = (-3)(-2)x\cdot x + 7x^2 = 6x^2 + 7x^2 = 13x^2$.

61 $(-x^2yz^4)(-6x^8y^4z^2) = (-1)(-6)\cdot(x^2x^8)\cdot(yy^4)\cdot(z^4z^2) = 6x^{10}y^5z^6$.

65 $\dfrac{66r^{12}s^9t^4}{11r^{13}s^3t^4} = \dfrac{66}{11}\cdot\dfrac{r^{12}}{r^{13}}\cdot\dfrac{s^9}{s^3}\cdot\dfrac{t^4}{t^4} = 6\cdot\dfrac{1}{r^1}\cdot s^6 \cdot 1 = \dfrac{6s^6}{r}$.

69 $(3x^4)^2 + 7x^8 = 3^2(x^4)^2 + 7x^8 = 9x^8 + 7x^8 = 16x^8$.

73 $\dfrac{(2xy)^3(-x^2y)^4}{(-x^3y)^7(8xy)^2} = \dfrac{2^3 x^3 y^3 \cdot (-1)^4 x^8 y^4}{(-1)^7 x^{21} y^7 \cdot 64 x^2 y^2} = \dfrac{8x^{11}y^7}{-64\,x^{23}y^9} = -\dfrac{1}{8x^{23-11}y^{9-7}} = -\dfrac{1}{8x^{12}y^2}$.

Exercises 3.2

[77] $3a^2b(6ab^2 - 4ab^2) = 3a^2b(2ab^2) = 6a^3b^3$.

[81] $1.73x - 2.86x + .0x = (1.73 - 2.86 + 0)x = -1.13x$.

Review Problems

[89] If $x = 2$ and $y = -3$ then $x^2 - y^2 = 2^2 - (-3)^2 = 4 - 9 = -5$.

[93] $3 + 2(6 - 4) = 3 + 2 \cdot 2 = 3 + 4 = 7$.

Exercises 3.3

[1] The degree of $3x - 1$ is one.

[5] The degree of $3a^4b^7 + a^2b^{20}$ is 22. Note that the degree of the first term is $4 + 7 = 11$ and the degree of the 2nd term is $2 + 20 = 22$. The degree of the polynomial is the degree of the term with highest degree.

[9] When $z = -1$, $z^{97} - 3z^{30} + 1 = (-1)^{97} - 3(-1)^{30} + 1 = -1 - 3 \cdot 1 + 1 = -1 - 3 + 1 = -3$.

[13] When $y = -4$, $2y^2 - 3y + 8 = 2(-4)^2 - 3(-4) + 8 = 2 \cdot 16 + 12 + 8 = 32 + 12 + 8 = 52$.

[17] When $x = -7$ and $y = -12$, $x^3 - 3xy^5 = (-7)^3 - 3(-7)(-12)^5 = -343 + 21(-248,832) = -343 - 5,225,472 = -5,225,815$.

[21] $(x^2 - 7x + 2) + (3x^2 + x - 8) = (x^2 + 3x^2) + (-7x + x) + (2 - 8) = 4x^2 - 6x - 6$.

Exercises 3.3

[25]
$3x^2 + 4xy - 2y^2$
$2x^2 - 3xy + y^2$
$\overline{5x^2 + xy - y^2}$

[29]
$12a^2 + 4ab + 12b^2$
$-a^2 + 12ab + 2b^2$
$6a^2 - ab + b^2$
$\overline{17a^2 + 15ab + 15b^2}$

[33] $(2a^2 - ab) - (ab - 3a^2) = (2a^2 - ab) + (-1)(ab - 3a^2) = 2a^2 - ab - ab + 3a^2 = 5a^2 - 2ab$.

[37] $(x^2 - 3x) - (2x^2 - x + 5) = x^2 - 3x - 2x^2 + x - 5 = -x^2 - 2x - 5$.

[41] $3(4x - 5) = 12x - 15$.

[45] $6x - 4x(3x - 2) = 6x - 4x(3x) - 4x(-2) = 6x - 12x^2 + 8x = -12x^2 + 14x$.

[49] $(3z - 7) + 2(6z + 4) = 3z - 7 + 2 \cdot 6z + 2 \cdot 4 = 3z - 7 + 12z + 8 = 15z + 1$.

[53] $-2m^4(4m^2 - 5m + 7) = -2m^4(4m^2) - 2m^4(-5m) - 2m^4(7) = -8m^6 + 10m^5 - 14m^4$.

[57] $(3x - 4)2x + (3x - 4)5 = 6x^2 - 8x + 15x - 20 = 6x^2 + 7x - 20$.

[61] $a(a^2 - 2a + 4) + 2(a^2 - 2a + 4) = a^3 - 2a^2 + 4a + 2a^2 - 4a + 8 = a^3 + 8$.

[65] $(a - 4) + (6a + 9) - (3a - 4) = a - 4 + 6a + 9 - 3a + 4 = 4a + 9$.

[69] $\left(\frac{1}{3}y - \frac{2}{5}\right) - \left(\frac{1}{4}y + \frac{1}{10}\right) = \frac{1}{3}y - \frac{2}{5} - \frac{1}{4}y - \frac{1}{10} = \left(\frac{1}{3} - \frac{1}{4}\right)y - \frac{2}{5} - \frac{1}{10} = \left(\frac{4}{12} - \frac{3}{12}\right)y - \frac{4}{10} - \frac{1}{10} = \frac{1}{12}y - \frac{5}{10} = \frac{1}{12}y - \frac{1}{2}$.

[73] $.2x(.1x^2 - 10) = (.2)(.1)x^3 - (.2)(10)x = .02x^3 - 2x$.

Review Problems

[81] If the side of a square is 6 cm then its area is $6^2 = 36$ cm^2.

Exercises 3.3

85 Let x, $x+2$, and $x+4$ be the 3 consecutive even integers. Then
$x + 4 = x + x + 2 + 10 \Rightarrow x + 4 = 2x + 12 \Rightarrow -x = 8 \Rightarrow x = -8$.
$x + 2 = -8 + 2 = -6$ and $x + 4 = -8 + 4 = -4$.
Therefore the even integers are -8, -6, and -4.

Exercises 3.4

1 $3z - 4$ is a linear binomial.

5 -8 is a constant. It has degree 0.

9 $2x^3 - 3x + 5$ is a cubic trinomial.

13 a) $(x+6)(x+1) = (x+6) \cdot x + (x+6) \cdot 1 = (x^2 + 6x) + (x+6) = x^2 + 7x + 6$.
b) $(x+6)(x+1) = (x \cdot x) + (x \cdot 1) + (6 \cdot x) + (6 \cdot 1) = x^2 + x + 6x + 6 = x^2 + 7x + 6$.

17 a) $(x+4)^2 = (x+4)(x+4) = (x+4) \cdot x + (x+4) \cdot 4 = x^2 + 4x + 4x + 16 = x^2 + 8x + 16$.
b) $(x+4)(x+4) = (x \cdot x) + (4 \cdot x) + (x \cdot 4) + (4 \cdot 4) = x^2 + 4x + 4x + 16 = x^2 + 8x + 16$.

21 a) $(x+4)(x+5) = x^2 + 5x + 4x + 20 = x^2 + 9x + 20$.
b)
$\quad x + 4$
$\quad x + 5$
$\overline{\quad x^2 + 4x \quad\quad\quad}$
$\quad\quad\quad 5x + 20$
$\overline{\quad x^2 + 9x + 20}$

25 a) $(x-2)(x^2 - 3x + 1) = x(x^2 - 3x + 1) - 2(x^2 - 3x + 1) = x^3 - 3x^2 + x - 2x^2 + 6x - 2 = x^3 - 5x^2 + 7x - 2$.

Exercises 3.4

b)
$$\begin{array}{r} x^2 - 3x + 1 \\ x - 2 \\ \hline x^3 - 3x^2 + x \\ - 2x^2 + 6x - 2 \\ \hline x^3 - 5x^2 + 7x - 2 \end{array}$$

29 $2x(x^2 - 7x + 3) = 2x \cdot x^2 + 2x \cdot (-7x) + 2x \cdot 3 = 2x^3 - 14x^2 + 6x$.

33 $2a(a+9)(a-9) = 2a(a^2 - 9a + 9a - 81) = 2a(a^2 - 81) = 2a \cdot a^2 - 2a \cdot 81 = 2a^3 - 162a$.

37 $(2x - 5y)(3x - 2y) = 6x^2 - 4xy - 15xy + 10y^2 = 6x^2 - 19xy + 10y^2$.

41
$$\begin{array}{r} x^2 + 2x + 4 \\ x - 2 \\ \hline x^3 + 2x^2 + 4x \\ - 2x^2 - 4x - 8 \\ \hline x^3 \phantom{{}-2x^2-4x} - 8 \end{array}$$

45 To expand $(2x-1)^3$, first expand $(2x-1)^2$:
$(2x-1)^2 = (2x-1)(2x-1) = 4x^2 - 2x - 2x + 1 = 4x^2 - 4x + 1$. Then note that $(2x-1)^3 = (2x-1)^2(2x-1) = (4x^2 - 4x + 1)(2x-1)$. Expand this vertically:

$$\begin{array}{r} 4x^2 - 4x + 1 \\ 2x - 1 \\ \hline 8x^3 - 8x^2 + 2x \\ - 4x^2 + 4x - 1 \\ \hline 8x^3 - 12x^2 + 6x - 1 \end{array}$$

Therefore $(2x-1)^3 = 8x^3 - 12x^2 + 6x - 1$.

49
$$\begin{array}{r} x^2 + x + 2 \\ x^2 + x + 2 \\ \hline x^4 + x^3 + 2x^2 \\ x^3 + x^2 + 2x \\ 2x^2 + 2x + 4 \\ \hline x^4 + 2x^3 + 5x^2 + 4x + 4 \end{array}$$

53
$8x - 2 = 5x + 19$
$8x - 2 + 2 = 5x + 19 + 2$
$8x = 5x + 21$
$8x - 5x = 5x + 21 - 5x$
$3x = 21$ or $x = \frac{21}{3} = 7$.

Exercises 3.4

57
$(x+4)(x-4) = x^2 - 8x$ ⇐ Remove the parentheses
$x^2 - 4x + 4x - 16 = x^2 - 8x$
$x^2 - 16 = x^2 - 8x$ ⇐ Subtract x^2 from both sides
$x^2 - 16 - x^2 = x^2 - 8x - x^2$
$-16 = -8x$ ⇐ Divide both sides by -8
$\frac{-16}{-8} = \frac{-8x}{-8}$ or $x = 2$.

61
$x(x+1) = (x+1)(x+2) - 16$ ⇐ Remove the parentheses
$x^2 + x = x^2 + 2x + x + 2 - 16$ ⇐ Combine like terms
$x^2 + x = x^2 + 3x - 14$ ⇐ Subtract x^2 from both sides
$x = 3x - 14$ ⇐ Subtract x from both sides and add 14 to both sides
$14 = 2x$ ⇐ Divide both sides by 2
$7 = x$ or $x = 7$.

65 Problem Statement: Find two consecutive even integers such that the sum of their squares is 28 less than twice the square of the larger integer.

Let x and $x+2$ be the two even integers. Then $x^2 + (x+2)^2 = 2(x+2)^2 - 28$.

To solve this equation, remove the grouping symbols and combine like terms:

$x^2 + x^2 + 4x + 4 = 2(x^2 + 4x + 4) - 28$
$2x^2 + 4x + 4 = 2x^2 + 8x + 8 - 28$ ⇐ Subtract $2x^2$ from both sides and simplify
$4x + 4 = 8x - 20$ ⇐ Subtract $4x$ from both sides and add 20 to both sides
$24 = 4x$ or $x = \frac{24}{4} = 6$. Therefore the integers are 6 and 8.

69 Problem Statement: The side of one square is 3 cm longer than the side of another square. The difference of their areas is 7 more than 4 times the length of the side of the larger square. Find the side of each square.

Let x and $x+3$ be the lengths of the sides of the 2 squares. The difference of their areas is given by $(x+3)^2 - x^2$ and 7 more than 4 times the length of the side of the larger square is $7 + 4(x+3)$. Therefore $(x+3)^2 - x^2 = 7 + 4(x+3)$. To solve, remove the grouping symbols:
$x^2 + 6x + 9 - x^2 = 7 + 4x + 12$ which simplifies to $6x + 9 = 4x + 19 \Rightarrow 2x = 10 \Rightarrow x = 5$.
Therefore the sides are 5 cm and 8 cm.

Exercises 3.4

73 Problem Statement: A circular pool has a radius of 8 m and is surrounded by a walk 2 m wide. Find the area of the walk.

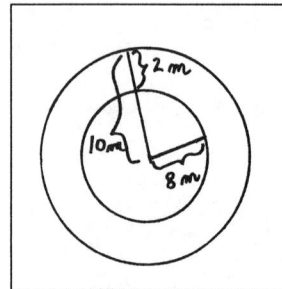

Since $A = \pi r^2$, the area of the pool is $\pi(8)^2 = 64\pi$ and the area of the pool plus the walk is $\pi(10)^2 = 100\pi$.
Therefore the area of the walk is $100\pi - 64\pi = 36\pi$.

77 Problem Statement: Find three consecutive even integers such that when the product of the first two is subtracted from the product of the last two, the result is 32.

Let x, $x+2$, and $x+4$ be the three even integers. Then the product of the 1st two is $x(x+2)$ and the product of the last two is $(x+2)(x+4)$. Therefore
$(x+2)(x+4) - x(x+2) = 32 \Rightarrow x^2 + 6x + 8 - x^2 - 2x = 32 \Rightarrow 4x + 8 = 32 \Rightarrow 4x = 24 \Rightarrow x = 6$.
Therefore the 3 integers are 6, 8, and 10.

81 To expand $(t+1)^4$, note that $(t+1)^4 = (t+1)^2(t+1)^2 = (t^2 + 2t + 1)(t^2 + 2t + 1)$.

Do this last product vertically:
$$
\begin{array}{r}
t^2 + 2t + 1 \\
t^2 + 2t + 1 \\
\hline
t^4 + 2t^3 + t^2 \\
2t^3 + 4t^2 + 2t \\
t^2 + 2t + 1 \\
\hline
t^4 + 4t^3 + 6t^2 + 4t + 1
\end{array}
$$

85 $(a^m + b^n)(a^m - b^n) = a^m a^m - a^m b^n + b^n a^m - b^n b^n = a^{m+m} - a^m b^n + a^m b^n - b^{n+n} = a^{2m} - b^{2n}$.

89 $(a+b)^2(a-b)^2 = [(a+b)(a-b)]^2 = [a^2 - b^2]^2 = (a^2 - b^2)(a^2 - b^2) = a^4 - 2a^2 b^2 + b^4$.

Review Problems

93 $\dfrac{4x^{10}}{2x^2} = 2x^{10-2} = 2x^8$.   **97** $(-3y+8)-(-5y+2) = -3y+8+5y-2 = 2y+6$.

Exercises 3.5

1 $\dfrac{6x^8}{3x^2} = 2x^{8-2} = 2x^6$.   **5** $\dfrac{12t^5 - 10t^2}{6t^3} = \dfrac{12t^5}{6t^3} - \dfrac{10t^2}{6t^3} = 2t^{5-3} - \dfrac{5 \cdot \cancel{2}}{3 \cdot \cancel{2} t^{3-2}} = 2t^2 - \dfrac{5}{3t}$.

9 $\dfrac{2m^8n^2 - 6mn^3 + 12n^4}{18m^2n^5} = \dfrac{2m^8n^2}{18m^2n^5} - \dfrac{6mn^3}{18m^2n^5} + \dfrac{12n^4}{18m^2n^5} =$

$\dfrac{\cancel{2}m^{8-2}}{9 \cdot \cancel{2} n^{5-2}} - \dfrac{\cancel{6} \cdot 1}{3 \cdot \cancel{6} m^{2-1} n^{5-3}} + \dfrac{\cancel{6} \cdot 2}{\cancel{6} \cdot 3 m^2 n^{5-4}} = \dfrac{m^6}{9n^3} - \dfrac{1}{3mn^2} + \dfrac{2}{3m^2n}$.

13 $\begin{array}{r} x^2 - 6x \\ -x^2 + 5x \\ \hline 2x^2 - 11x \end{array}$ To do this subtraction note that $x^2 - (-x^2) = x^2 + x^2 = 2x^2$ and $-6x - 5x = -11x$.

17 $\begin{array}{r} z^2 + 0z - 8 \\ z^2 + 3z - 2 \\ \hline -3z - 6 \end{array}$ Note that $z^2 - z^2 = 0$; $0z - 3z = -3z$; and $-8 - (-2) = -8 + 2 = -6$.

25 Write $4x^2 - 3$ as $4x^2 + 0x - 3$ so that a place is held for the x terms.

$\begin{array}{r} 4x - 20 \\ x+5 \overline{\smash{\big)}\, 4x^2 + 0x - 3} \\ \underline{4x^2 + 20x} \\ -20x - 3 \\ \underline{-20x - 100} \\ 97 \end{array}$ Therefore $\dfrac{4x^2 - 3}{x+5} = 4x - 20 + \dfrac{97}{x+5}$.

Exercises 3.5

29 When doing the long division, rewrite $4x^2 + 3 - 8x$ as $4x^2 - 8x + 3$.

$$\begin{array}{r} 2x - 1 \\ 2x-3 \overline{\smash{\big)}\, 4x^2 - 8x + 3} \\ \underline{4x^2 - 6x} \\ -2x + 3 \\ \underline{-2x + 3} \\ 0 \end{array}$$

Therefore $\dfrac{4x^2 - 8x + 3}{2x - 3} = 2x - 1$

33 $\dfrac{12x^2 + 5x - 3}{3x} = \dfrac{12x^2}{3x} + \dfrac{5\not{x}}{3\not{x}} - \dfrac{\not{3}\cdot 1}{\not{3}x} = 4x + \dfrac{5}{3} - \dfrac{1}{x}$.

37 $\dfrac{10m^3 - 15m^2 + 20}{10m^3} = \dfrac{10m^3}{10m^3} - \dfrac{15m^2}{10m^3} + \dfrac{20}{10m^3} = 1 - \dfrac{\not{5}\cdot 3}{\not{5}\cdot 2m^{3-2}} + \dfrac{2 \cdot 1\not{0}}{1\not{0}m^3} = 1 - \dfrac{3}{2m} + \dfrac{2}{m^3}$.

41

$$\begin{array}{r} x + 2 \\ x-2 \overline{\smash{\big)}\, x^2 + 0x + 1} \\ \underline{x^2 - 2x} \\ 2x + 1 \\ \underline{2x - 4} \\ 5 \end{array}$$

Therefore $\dfrac{x^2 + 1}{x - 2} = x + 2 + \dfrac{5}{x - 2}$.

45 Rewrite $27 + a^3$ as $a^3 + 0a^2 + 0a + 27$.

$$\begin{array}{r} a^2 - 3a + 9 \\ a+3 \overline{\smash{\big)}\, a^3 + 0a^2 + 0a + 27} \\ \underline{a^3 + 3a^2} \\ -3a^2 + 0a + 27 \\ \underline{-3a^2 - 9a} \\ 9a + 27 \\ \underline{9a + 27} \\ 0 \end{array}$$

Therefore $\dfrac{27 + a^3}{a + 3} = a^2 - 3a + 9$.

Exercises 3.5

49 Rewrite $3y - 3 + 6y^3 - 8y^2$ as $6y^3 - 8y^2 + 3y - 3$.

$$\begin{array}{r} 3y - 4 \\ 2y^2 + 1 \overline{\smash{)}6y^3 - 8y^2 + 3y - 3} \\ \underline{6y^3 + 3y } \\ -8y^2 - 3 \\ \underline{-8y^2 - 4} \\ 1 \end{array}$$

Therefore $\dfrac{3y - 3 + 6y^3 - 8y^2}{2y^2 + 1} = 3y - 4 + \dfrac{1}{2y^2 + 1}$.

53 To simplify $\dfrac{x^2 - 1}{x + 1} \cdot 3x + 4x$, do the division $\dfrac{x^2 - 1}{x + 1}$ first:

$$\begin{array}{r} x - 1 \\ x + 1 \overline{\smash{)}x^2 + 0x - 1} \\ \underline{x^2 + x } \\ -x - 1 \\ \underline{-x - 1} \\ 0 \end{array}$$

Therefore $\dfrac{x^2 - 1}{x + 1} = x - 1$ and $\dfrac{x^2 - 1}{x + 1} \cdot 3x + 4x = (x - 1)3x + 4x$.

Now multiply $(x - 1)$ by $3x$:

$(x - 1)3x + 4x = 3x^2 - 3x + 4x = 3x^2 + x$.

Ans: $3x^2 + x$

57 Rewrite $2x - 1 + x^2$ as $x^2 + 2x - 1$.

$$\begin{array}{r} x^2 - 2x - 3 \\ x^2 + 2x - 1 \overline{\smash{)}x^4 + 0x^3 - 8x^2 + 3x + 2} \\ \underline{x^4 + 2x^3 - x^2 } \\ -2x^3 - 7x^2 + 3x + 2 \\ \underline{-2x^3 - 4x^2 + 2x } \\ -3x^2 + x + 2 \\ \underline{-3x^2 - 6x + 3} \\ 7x - 1 \end{array}$$

Therefore $\dfrac{x^4 - 8x^2 + 3x + 2}{x^2 + 2x - 1} =$

$x^2 - 2x - 3 + \dfrac{7x - 1}{x^2 + 2x - 1}$.

61

$$\begin{array}{r} 3x^2 + 2 \\ 3x^2 + 2 \overline{\smash{)}9x^4 + 12x^2 + 4} \\ \underline{9x^4 + 6x^2 } \\ 6x^2 + 4 \\ \underline{6x^2 + 4} \\ 0 \end{array}$$

Therefore $\dfrac{9x^4 + 12x^2 + 4}{3x^2 + 2} = 3x^2 + 2$.

Exercises 3.5

65. $\dfrac{a^{12}b^3c^{10} - a^2bc + b^{80}c^{90}}{a^2bc^3} = \dfrac{a^{12}b^3c^{10}}{a^2bc^3} - \dfrac{a^2bc}{a^2bc^3} + \dfrac{b^{80}c^{90}}{a^2bc^3} =$

$= \dfrac{a^{12}}{a^2} \cdot \dfrac{b^3}{b} \cdot \dfrac{c^{10}}{c^3} - \dfrac{a^2b}{a^2b} \cdot \dfrac{c}{c^3} + \dfrac{1}{a^2} \cdot \dfrac{b^{80}}{b} \cdot \dfrac{c^{90}}{c^3} = a^{10}b^2c^7 - \dfrac{1}{c^2} + \dfrac{b^{79}c^{87}}{a^2}$.

Review Exercises

1. $9x^2$ is a quadratic monomial.

5. $4x^2 - 7x + 6$ is a quadratic trinomial.

9. $t^2(5t^3)^3 = t^2 \cdot 5^3 \cdot (t^3)^3 = t^2 \cdot 125 \cdot t^9 = 125t^{11}$.

13. $\left(\dfrac{a^4}{6}\right)^2 = \dfrac{(a^4)^2}{6^2} = \dfrac{a^8}{36}$.

17. $3a^2b + 2ab^2 - a^2b = (3a^2b - a^2b) + 2ab^2 = 2a^2b + 2ab^2$.

21. $(-2a^4b^6)^3 = (-2^3)(a^4)^3(b^6)^3 = -8a^{12}b^{18}$.

25. If $a = 2$ and $b = -3$ then $9a^2 - 4b^2 = 9 \cdot (2)^2 - 4 \cdot (-3)^2 = 9 \cdot 4 - 4 \cdot 9 = 36 - 36 = 0$.

29. $\dfrac{2}{3}(12x + 15y) = \dfrac{2}{3} \cdot 12x + \dfrac{2}{3} \cdot 15y = 2 \cdot 4x + 2 \cdot 5y = 8x + 10y$.

33. $(3x - 2)(4x + 5) = 12x^2 + 15x - 8x - 10 = 12x^2 + 7x - 10$.

37. To solve $x(3x - 4) + 1 = 3x(x + 6)$ remove the grouping symbols. When this is done we get $3x^2 - 4x + 1 = 3x^2 + 18x$ which simplifies to $1 = 22x$ which implies that $x = \dfrac{1}{22}$.

41. Let x and $x + 2$ be the integers. Then $(x+2)^2 - x^2 = 36$ which is equivalent to $x^2 + 4x + 4 - x^2 = 36$ which reduces to $4x = 32$ or $x = \dfrac{32}{4} = 8$. Therefore the integers are 8 and 10.

45. $\dfrac{6x^2 + 7x - 4}{2x} = \dfrac{6x^2}{2x} + \dfrac{7x}{2x} - \dfrac{4}{2x} = 3x + \dfrac{7}{2} - \dfrac{2}{x}$.

Chapter 3 Review Exercises and Test

49 Rewrite $3y + 4 + 3y^2 + y^3$ so that the powers are in ascending order.

$$\begin{array}{r} y^2 + 2y + 1 \\ y+1 \overline{\smash{\big)}\, y^3 + 3y^2 + 3y + 4} \\ \underline{y^3 + y^2} \\ 2y^2 + 3y \\ \underline{2y^2 + 2y} \\ y + 4 \\ \underline{y + 1} \\ 3 \end{array}$$

Therefore $\dfrac{3y + 4 + 3y^2 + y^3}{y+1} = y^2 + 2y + 1 + \dfrac{3}{y+1}$.

Chapter Test

1. $2x^3 - x + 6$ is a cubic trinomial.

5. $(x+3)^2 = (x-4)^2$ is equivalent to $x^2 + 6x + 9 = x^2 - 8x + 16$ which reduces to $14x = 7$ which implies that $x = \dfrac{7}{14} = \dfrac{1}{2}$.

9. $a^6 a^4 = a^{6+4} = a^{10}$.

13. $\left(\dfrac{3x^4 y^2}{9xy^3}\right)^2 = \dfrac{9x^8 y^4}{81 x^2 y^6} = \dfrac{x^6}{9y^2}$.

17. $x^2 - (6x^2 - 3x + 4) = x^2 - 6x^2 + 3x - 4 = -5x^2 + 3x - 4$.

21. $(a - 2b)^3 = (a - 2b)^2 (a - 2b) = (a^2 - 4ab + 4b^2)(a - 2b) =$
$a^3 - 4a^2 b + 4ab^2 - 2a^2 b + 8ab^2 - 8b^3 = a^3 - 6a^2 b + 12ab^2 - 8b^3$.

25.
$$\begin{array}{r} 2y - 3 \\ 2y^2 + 1 \overline{\smash{\big)}\, 4y^3 - 6y^2 + 0y + 7} \\ \underline{4y^3 + 2y} \\ -6y^2 - 2y + 7 \\ \underline{-6y^2 - 3} \\ -2y + 10 \end{array}$$

Therefore $\dfrac{4y^3 - 6y^2 + 7}{2y^2 + 1} = 2y - 3 + \dfrac{-2y + 10}{2y^2 + 1}$.

Cumulative Review: Chapters 1-3

Cumulative Review

1. $3^2 + 20 \div 5 \cdot 4 = 9 + 4 \cdot 4 = 9 + 16 = 25$.

5. $2(3x - 4) - (8 - 7x) = 6x - 8 - 8 + 7x = 13x - 16$.

9. $(3x + 5)^2 = (3x)^2 + 2(3x)(5) + 25 = 9x^2 + 30x + 25$.

13. $x^4 \cdot x^5 = x^{4+5} = x^9$.

17. $(5a^4)^3 = 5^3(a^4)^3 = 125a^{12}$.

21. $2x - 1 < 5$ is equivalent to $2x < 6$ which implies that $x < \frac{6}{2}$ or $x < 3$.

```
◄─────────○
  0  1  2  3  4  5
```

25. $0x = 0$ is true for all real numbers.

29. $P = 2L + 2W$
    $P - 2W = 2L$
    $L = \frac{P - 2W}{2}$.

33. Let $x = .\overline{23}$. Then $100x = 23.\overline{23}$ and
    $100x - x = 23$ which implies that
    $99x = 23$ or $x = \frac{23}{99}$.

37. The irrational numbers are π and $.81828384858\ldots$.

41. If $x = -2$, $y = -1$, and $z = 3$ then $2xy - 5z = 2(-2)(-1) - 5(3) = 4 - 15 = -11$.

45. Let x be the amount of 15% salt solution to be mixed. Then the mixture totals $x + 6$ and the amount of salt in the mixture is $.20(8) + .15x$ or $1.6 + .15x$. Since the mixture must be 17% salt, the following equation must be true:
    $1.6 + .15x = .17(x + 8) \Rightarrow 1.6 + .15x = .17x + 1.36$. If we multiply both sides by 100 we get
    $160 + 15x = 17x + 136 \Rightarrow 24 = 2x \Rightarrow x = \frac{24}{2} = 12$.
    Ans: Add 12 gallons of 15% salt solution.

49. Let x be the number of quarters. Then $16 - x$ is the number of dimes. The quarters are worth $25x$ cents and the dimes are worth $10(16 - x)$ cents. Therefore, since the total is worth \$3.40, we have the following equation:

$25x + 10(16 - x) = 340 \Rightarrow 25x + 160 - 10x = 340$ which simplifies to $15x = 180$ or $x = \frac{180}{15} = 12$. Ans: 12 quarters and 4 dimes.

Chapter 4: Factoring

|Exercises 4.1|

[1]   28 is a composite number because $28 = 2 \times 2 \times 7$.

[5]   0 is neither because prime numbers and composite numbers must be greater than 1.

[9]   The positive factors of 6 are 1, 2, 3, and 6.

[13]   35 is not divisible by 2 since its last digit is not divisible by 2. It is not divisible by 3 since the sum of the digits, 8, is not divisible by 3. It is not divisible by 9 because the sum of the digits is not divisible by 9. However it is divisible by 5 since the last digit is 5. Therefore 5 is a factor and $35 = 5 \times 7$. So we see that 7 is also a factor.

[17]   190 is divisible by 2, 5, and 10 since the last digit is 0. Hence $190 = 2 \times 5 \times 19$, and we see that 19 is also a factor.

[21]   The prime factorization of 35 is 5×7.   [25] $54 = 2 \times 27 = 2 \times 3^3$

[29] $125 = 5 \times 25 = 5 \times 5 \times 5 = 5^3$.

[33]   Since $8 = 2^3$ and $12 = 2^2 \times 3$ we see that 8 and 12 are both divisible by 2, 4, and of course 1. Therefore the common factors are 1, 2, and 4.

[37]   Since $21 = 3 \times 7$ and $27 = 3^3$ we see that the largest integer that divides both 21 and 27 is 3. Therefore the GCF of (21, 27) is 3.

[41]   Since $18 = 6 \times 3$, $30 = 6 \times 5$, and $42 = 6 \times 7$ we see that the largest common factor of 18, 30, and 42 is 6. Therefore the GCF is 6.

Exercises 4.1

[45] To complete the factoring divide $-24m^{18}$ by $12m^4$: $\dfrac{-24m^{18}}{12m^4} = -2m^{14}$. Therefore $-24m^{18} = 12m^4\left[-2m^{14}\right]$.

[49] Since $8x^2 = 4x \cdot 2x$ and $12x = 4x \cdot 3$ we see that the GCF is $4x$.

[53] Since $125a^3b = 5 \cdot 25 \cdot \mathbf{ab}(a^2)$ and $-45ab^3 = 5(-9)\mathbf{ab}(b^2)$ we see that the GCF is $5ab$.

[57] $10 - 30t^2 = \mathbf{10} \cdot 1 - \mathbf{10} \cdot 3t^2 = 10(1 - 3t^2)$.

[61] $12y^4 + 4y^2 = \mathbf{4y^2}(3y^2) + \mathbf{4y^2} \cdot 1 = 4y^2(3y^2 + 1)$.

[65] $8ab + 3ab + 15cd - 4cd = 11ab + 11cd = 11(ab + cd)$.

[69] $6a^6 - 5a^4 - 3a^3 = \mathbf{a^3} \cdot 6a^3 - \mathbf{a^3} \cdot 5a - \mathbf{a^3} \cdot 3 = a^3(6a^3 - 5a - 3)$.

[73] $6a^2 - 9b^2 + 21v^2 = 3 \cdot 2a^2 - 3 \cdot 3b^2 + 3 \cdot 7v^2 = 3(2a^2 - 3b^2 + 7v^2)$.

[77] $3a^3b - 6a^2b^2 + 3ab^3 = \mathbf{3ab} \cdot a^2 - \mathbf{3ab} \cdot 2ab + \mathbf{3ab} \cdot b^2 = 3ab(a^2 - 2ab + b^2)$.
Note that later on you will see that $a^2 - 2ab + b^2 = (a-b)(a-b)$.

[81] $2x^2 + 6x + 2 = 2(x^2 + 3x + 1)$ is true because $2(x^2 + 3x + 1) = 2 \cdot x^2 + 2 \cdot 3x + 2 \cdot 1 = 2x^2 + 6x + 2$.

[85] It is not completely factored. $6x^4 + 12x^3 - 6x^2 = 6x^2(x^2 + 2x - 1)$.

[89] $128x^{30} - 192x^{83} = \mathbf{64} \cdot 2 \cdot \mathbf{x^{30}} - \mathbf{64} \cdot 3 \cdot \mathbf{x^{30}} \cdot x^{53} = 64x^{30}(2 - 3x^{53})$.

[93] To factor $a^2bx + ab^2x + 3a^2b + 3ab^2$ note that each term is divisible by ab. Therefore $a^2bx + ab^2x + 3a^2b + 3ab^2 = ab(ax + bx + 3a + 3b)$. However the factorization is not complete. $ab(ax + bx + 3a + 3b) = ab[x(\mathbf{a+b}) + 3(\mathbf{a+b})] = ab[(a+b)(x+3)] = ab(a+b)(x+3)$.

Exercises 4.1

[97] $(4x+6)(3x-9) = 12x^2 - 36x + 18x - 54 = 12x^2 - 18x - 54$. Note that each term is divisible by 6. This is the largest such number. Therefore the GCF is 6. A better way to do this is to factor $4x+6$ and $3x-9$: $(4x+6)(3x-9) = 2\cdot(2x+3)3\cdot(x-3) = 6(2x+3)(x-3)$.

[101] The divisors of 6 are 1, 2, 3, and 6. The reciprocals of the divisors are $\frac{1}{1} = 1, \frac{1}{2}, \frac{1}{3}$, and $\frac{1}{6}$. $1 + \frac{1}{2} + \frac{1}{3} + \frac{1}{6} = \frac{6}{6} + \frac{3}{6} + \frac{2}{6} + \frac{1}{6} = \frac{12}{6} = 2$. Therefore 6 is a perfect number.

Review Problems

[105] $(a+3)(a-3) = a^2 - 3a + 3a - 9 = a^2 - 9$.

[109] $x^2 + 3x + 15 = x^2 - 2x$ reduces to $5x = -15$ which implies that $x = \frac{-15}{5} = -3$.

Exercises 4.2

[1] $x^2 - 25 = x^2 - 5^2 = (x+5)(x-5)$.
[5] $9 - t^2 = 3^2 - t^2 = (3+t)(3-t)$.

[9] $(a-3)^2 - b^2 = [(a-3)+b][(a-3)-b] = (a-3+b)(a-3-b)$.

[13] $7x^2 - 28 = 7(x^2 - 4) = 7(x^2 - 2^2) = 7(x+2)(x-2)$.

[17] $9x^2 + 36 = 9(x^2 + 4)$.
[21] $64x^2 - 49a^2b^2 = (8x)^2 - (7ab)^2 = (8x+7ab)(8x-7ab)$.

[25] $(x-3)(x+7) = 0$ when $x-3 = 0$ or $x+7 = 0$ which implies that $x = 3$ or -7.

Exercises 4.2

29 $x^2 - 25 = 0$
$(x+5)(x-5) = 0$
$x + 5 = 0$ or $x - 5 = 0$
$x = -5$ or $x = 5$.

33 $9x^2 - 16 = 0$
$(3x)^2 - 4^2 = 0$
$(3x+4)(3x-4) = 0$
$3x + 4 = 0$ or $3x - 4 = 0$
$3x = -4$ or $3x = 4$
$x = -\frac{4}{3}$ or $x = \frac{4}{3}$.

37 $4m^2 = m$
$4m^2 - m = 0$
$m(4m - 1) = 0$
$m = 0$ or $4m - 1 = 0$
$m = 0$ or $4m = 1$
$m = 0$ or $m = \frac{1}{4}$.

41 $16p^3 = 100p$
$16p^3 - 100p = 0$
$4p(4p^2 - 25) = 0$
$4p(2p + 5)(2p - 5) = 0$
$4p = 0$, $2p + 5 = 0$, or $2p - 5 = 0$
$p = 0$, $-\frac{5}{2}$, or $\frac{5}{2}$.

45 $(x+3)^2 = x^2 - 9$
$x^2 + 6x + 9 = x^2 - 9$
$6x = -18$
$x = \frac{-18}{6} = -3$.

49 $6x(2x-5)(2x+7) = 0$
$6x = 0$, $2x - 5 = 0$, or $2x + 7 = 0$
$x = \frac{0}{6} = 0$, $2x = 5$, or $2x = -7$
$x = 0, \frac{5}{2},$ or $-\frac{7}{2}$.

53 $x^3 + 8 = (x+2)^3$ is false. To factor $x^3 + 8$ use the fact that $a^3 + b^3 = (a+b)(a^2 - ab + b^2)$ where a is replaced by x and b is replaced by 2. Therefore
$x^3 + 8 = x^3 + 2^3 = (x+2)(x^2 - x \cdot 2 + 2^2) = (x+2)(x^2 - 2x + 4)$.

57 $x^3 + 1 = (x+1)(x^2 + 2x + 1)$ is false; $x^3 + 1 = x^3 + 1^3 = (x+1)(x^2 - x + 1)$.

61 To factor $m^3 - 4^3$, use the fact that $a^3 - b^3 = (a-b)(a^2 + ab + b^2)$ where $a = m$ and $b = 4$. Therefore $m^3 - 4^3 = (m-4)(m^2 + m \cdot 4 + 4^2) = (m-4)(m^2 + 4m + 16)$.

65 $8 - x^3 = 2^3 - x^3 = (2-x)(2^2 + 2 \cdot x + x^2) = (2-x)(4 + 2x + x^2)$.

69 $8m^3 - 27n^3 = (2m)^3 - (3n)^3 = (2m - 3n)[(2m)^2 + (2m)(3n) + (3n)^2] = (2m - 3n)(4m^2 + 6mn + 9n^2)$.

Exercises 4.2

[73] $27 + y^6 = 3^3 + (y^2)^3 = (3 + y^2)[3^2 - 3 \cdot y^2 + (y^2)^2] = (3 + y^2)(9 - 3y^2 + y^4)$.

[77] $(x + 2)^3 - y^3 = [(x + 2) - y][(x + 2)^2 + (x + 2) \cdot y + y^2] =$
$(x + 2 - y)[(x^2 + 4x + 4) + (xy + 2y) + y^2] = (x + 2 - y)(x^2 + 4x + 4 + xy + 2y + y^2)$.

[81] To factor $(x - 1)^2 - (x + 2)^2$, use the fact that $A^2 - B^2 = (A + B)(A - B)$ where A is replaced by $x - 1$ and B is replaced by $x + 2$. Therefore $(x - 1)^2 - (x + 2)^2 =$
$[(x - 1) + (x + 2)][(x - 1) - (x + 2)] = (x - 1 + x + 2)(x - 1 - x - 2) = (2x + 1) \cdot (-3) =$
$-3(2x + 1)$.

Review Problems

[89] $(k^3 + 2)(k + 4) = k^4 + 4k^3 + 2k + 8$.

Exercises 4.3

[1] To factor $x(x - 4) + 8(x - 4)$, note that $x - 4$ is a common factor. Therefore
$x(\mathbf{x - 4}) + 8(\mathbf{x - 4}) = (x - 4)(x + 8)$.

[5] $3k - 1$ is a common factor. Therefore $(\mathbf{3k - 1})2k + (\mathbf{3k - 1})5 = (3k - 1)(2k + 5)$.

[9] $c + d$ is a common factor. Therefore $a(\mathbf{c + d}) + b(\mathbf{c + d}) = (c + d)(a + b)$.

[13] To factor $y^2 + 7y + 7y + 49$, factor y from $y^2 + 7y$ and 7 from $7y + 49$ as follows:
$y^2 + 7y + 7y + 49 = y(y + 7) + 7(y + 7)$. Now note that $y + 7$ is a common factor and
$y(\mathbf{y + 7}) + 7(\mathbf{y + 7}) = (y + 7)(y + 7)$.

Exercises 4.3

17 To factor $4x^2 - 20x - x + 5$, factor $4x$ from $4x^2 - 20x$ and -1 from $-x + 5$ as follows:
$4x^2 - 20x - x + 5 = 4x(x-5) - 1(x-5)$. Now $x - 5$ is a common factor and
$4x(\mathbf{x-5}) - 1(\mathbf{x-5}) = (x-5)(4x-1)$.

21 $x^2 + 18 + 6x + 3x = (x^2 + 6x) + (3x + 18) = x(x+6) + 3(x+6) = (x+6)(x+3)$.

25 $x^6 + y^4 + y^2x^3 + x^3y^2 = (x^6 + x^3y^2) + (x^3y^2 + y^4) = x^3(x^3 + y^2) + y^2(x^3 + y^2) = (x^3 + y^2)(x^3 + y^2)$.

29 To factor $x^2 + kx - k - 1$, rewrite as $x^2 - 1 + kx - k$ and factor $x^2 - 1$ and factor $kx - k$.
$x^2 - 1 + kx - k = (x+1)(x-1) + k(x-1)$. Now $x - 1$ is a common factor and
$(x+1)(\mathbf{x-1}) + k(\mathbf{x-1}) = (x-1)(x+1+k)$.

33 $x - y$ is a common factor. Therefore $4(x-y) + (x+y)(x-y) = (x-y)(4+x+y)$.

37 To factor $2m^2 - 2mn + 2km^2 - 2kmn$, first factor out $2m$. Then
$2m^2 - 2mn + 2km^2 - 2kmn = 2m(m - n + km - kn)$. This can be factored further by factoring out the k from $km - kn$ and writing $m - n$ as $1 \cdot (m-n)$ as follows:
$2m(m - n + km - kn) = 2m[1 \cdot (\mathbf{m-n}) + k(\mathbf{m-n})] = 2m(m-n)(1+k)$.

41 $x^3 + 2x^2 - 25x - 50 = x^2(x+2) - 25(x+2) = (x+2)(x^2 - 25) = (x+2)(x+5)(x-5)$.

45 $x^2 - 2x - x + 2 = x(x-2) - 1(x-2) = (x-2)(x-1)$.

49 $x^2 + 4x + 2x + 10$ is prime.

53 $6y^2 - 10 + 4y - 15y = 6y^2 - 15y + 4y - 10 = 3y(2y-5) + 2(2y-5) = (2y-5)(3y+2)$.

57 $3x^2z - 12zy^2 = 3z(x^2 - 4y^2) = 3z(x+2y)(x-2y)$.

61 To factor $25a - 25b + a^2 - b^2$ first factor 25 from $25a - 25b$ and factor $a^2 - b^2$. Then
$25a - 25b + a^2 - b^2 = 25(\mathbf{a-b}) + (a+b)(\mathbf{a-b}) = (a-b)(25+a+b)$.

Exercises 4.3

65 $3a^2 - 3b^2 + (a-b)^2 = 3(a^2 - b^2) + (a-b)^2 = 3(a+b)(a-b) + (a-b)^2$. Now $a - b$ is a common factor and $3(a+b)(a-b) + (a-b)(a-b) = (a-b)[3(a+b) + (a-b)] = (a-b)(3a + 3b + a - b) = (a-b)(4a + 2b)$. Now 2 can be factored from $4a + 2b$. Thus $(a-b)(4a+2b) = (a-b) \cdot 2 \cdot (2a+b) = 2(a-b)(2a+b)$.

69 $a(t+5) - 3(t+5) + a(k+p) - 3(k+p) = (t+5)(a-3) + (k+p)(a-3) = (a-3)(t+5+k+p)$.

Review Problems

73 $(x+2)(x+3) = x^2 + 3x + 2x + 6 = x^2 + 5x + 6$.

77 $(x-10)(x+2) = 0$ if $x - 10 = 0$ or $x + 2 = 0$. Therefore $x = 10$ or -2.

Exercises 4.4

1 To factor $x^2 + 7x + 10$ into $(x+a)(x+b)$, a and b must satisfy the conditions that $ab = 10$ and $a + b = 7$. $a = 5$ and $b = 2$ satisfy these conditions since $5 \times 2 = 10$ and $5 + 2 = 7$. Therefore $x^2 + 7x + 10 = (x+5)(x+2)$.

5 To factor $x^2 - 18x + 81$ into $(x-a)(x-b)$, we must find a and b such that $ab = 81$ and $a + b = 18$. $a = 9$ and $b = 9$ satisfy these conditions since $9 \times 9 = 81$ and $9 + 9 = 18$. Therefore $x^2 - 18x + 81 = (x-9)(x-9)$.

9 $a^2 + 8ab + 16b^2$ can factor into $(a + mb)(a + nb)$ if we can find m and n such that $mn = 16$ and $m + n = 8$. $m = 4$ and $n = 4$ satisfy these conditions since $4 \times 4 = 16$ and $4 + 4 = 8$. Therefore $a^2 + 8ab + 16b^2 = (a + 4b)(a + 4b)$.

Exercises 4.4

[13] To factor $y^2 + 15y + 35$, there must exist positive integers a and b such that $ab = 35$ and $a + b = 15$. The only positive integers a and b such that $ab = 35$ are 5 and 7 or 35 and 1 and $5 + 7 = 12$ and $35 + 1 = 36$. Therefore $y^2 + 15y + 35$ is prime.

[17] To factor $z^3 + 9z^2 - 10z$ first factor out the z which gives $z(z^2 + 9z - 10)$. Now $z^2 + 9z - 10$ can be factored into $(z + a)(z + b)$ where a and b are integers such that $ab = -10$ and $a + b = 9$. Note that a and b must have opposite signs since their product is negative. $a = 10$ and $b = -1$ satisfy the conditions. Therefore $z(z^2 + 9z - 10) = z(z + 10)(z - 1)$.

[21] $x^2 + 11x + 28$ can factor into $(x + a)(x + b)$ if we can find positive integers a and b such that $ab = 28$ and $a + b = 11$. Therefore $a = 7$ and $b = 4$ and $x^2 + 11x + 28 = (x + 7)(x + 4)$.

[25] $3r^3 + 12r^2s - 12r^2s = 3r(r^2 + 4rs - 4s^2)$. Note that $r^2 + 4rs - 4s^2$ is prime.

[29] $2x^2 - 6x + 14$ has a common factor of 2. Therefore $2x^2 - 6x + 14 = 2(x^2 - 3x + 7)$. Note that $x^2 - 3x + 7$ is prime.

[33] $2x^3y - 3x^2z^2 + 24y^2z$ is prime.

[37] $x^2 + 10x + 21 = 0$
$(x + 7)(x + 3) = 0$
$x + 7 = 0$ or $x + 3 = 0$
$x = -7$ or $x = -3$.

[41] $20k + 100 = -k^2$
$k^2 + 20k + 100 = 0$
$(k + 10)(k + 10) = 0$
$k + 10 = 0$ or $k + 10 = 0$
Therefore $k = -10$ is the only solution.

[45] $x^2 = 7x$
$x^2 - 7x = 0$
$x(x - 7) = 0$
$x = 0$ or $x - 7 = 0$
$x = 0$ or $x = 7$.

[49] $18p^2 + 3p^3 = 120p$
$3p^3 + 18p^2 - 120p = 0$
$3p(p^2 + 6p - 40) = 0$
$3p(p + 10)(p - 4) = 0$
$p = 0, -10,$ or 4.

Exercises 4.4

53 $x^2 - 15xy + 54y^2$ can be factored into $(x - ay)(x - by)$ where $ab = 54$ and $a + b = 15$.
Therefore make $a = 9$ and $b = 6$ and $x^2 - 15xy + 54y^2 = (x - 9y)(x - 6y)$.

57 $a^2 + 34ab - 72b^2$ can be factored into $(a + mb)(a + nb)$ where $mn = -72$ and $m + n = 34$.
Note that one of these numbers must be positive and the other negative. If $m = 36$ and
$n = -2$ then $36 \times (-2) = -72$ and $36 + (-2) = 34$. Therefore
$a^2 + 34ab - 72b^2 = (a + 36b)(a - 2b)$.

61 $1 - 81x^4 = 1 - (9x^2)^2 = (1 + 9x^2)(1 - 9x^2) = (1 + 9x^2)(1 + 3x)(1 - 3x)$.

65 $x^2 - y^2 + 7x + 7y = (x + y)(x - y) + 7(x + y) = (x + y)(x - y + 7)$.

69 $x^2 + ax - 8a - 17x + 72 = (x^2 - 17x + 72) + (ax - 8a) = (x - 9)(x - 8) + a(x - 8) =$
$(x - 8)(x - 9 + a)$.

73 $2a^4 - 16a = 2a(a^3 - 8) = 2a(a^3 - 2^3) = 2a(a - 2)(a^2 + 2a + 4)$.

Review Problems

77 $(3x + 1)(x + 2) = 3x^2 + 6x + x + 2 = 3x^2 + 7x + 2$.

Exercises 4.5

1 $2x^2 + 9x + 4$ can be factored into $(ax + b)(cx + d)$ where $ac = 2$ and $bd = 4$. The possibilities
are $(2x + 1)(x + 4)$, $(2x + 4)(x + 1)$, or $(2x + 2)(x + 2)$. Note that 2 is not a common factor of
$2x^2 + 9x + 4$. However 2 is a common factor of $(2x + 4)(x + 1) = 2(x + 2)(x + 1)$.
This means that the factorization can not be $(2x + 4)(x + 1) = 2(x + 2)(x + 1)$.
Similarly the factorization can not be $(2x + 2)(x + 2)$. The correct factorization of $2x^2 + 9x + 4$
is $(2x + 1)(x + 4)$ which gives the correct middle term $2x \cdot 4 + 1 \cdot x = 8x + 1x = 9x$.

Exercises 4.5

[5] $8x^2 - 2xy - 3y^2$ can be factored into $(ax+b)(cx+d)$ where $ac = 8$ and $bd = -3$. The correct values are $a = 4$, $c = 2$, $b = -3$ and $d = 1$ because $4 \times 2 = 8$ and $-3 \times 1 = -3$ and we get the correct middle term: $(4x-3y)(2x+y) = 8x^2 + \mathbf{4xy - 6xy} - 3y^2 = 8x^2 - 2xy - 3y^2$. Therefore $8x^2 - 2xy - 3y^2 = (4x-3y)(2x+y)$.

[9] To factor $4x^2 + 20x + 25$, multiply $4 \times 25 = 100$ and find 2 numbers m and n such that $mn = 100$ and $m + n = 20$. The correct values are $m = 10$ and $n = 10$. Then $4x^2 + 20x + 25$ can be factored by grouping as follows: $4x^2 + 20x + 25 = 4x^2 + 10x + 10x + 25 = 2x(\mathbf{2x+5}) + 5(\mathbf{2x+5}) = (2x+5)(2x+5)$.

[13] Multiply $3 \times 2 = 6$ and find m and n such that $mn = 6$ and $m + n = 5$. The correct values are $m = 3$ and $n = 2$. Then $3x^2 + 5x + 2 = 3x^2 + 3x + 2x + 2 = 3x(\mathbf{x+1}) + 2(\mathbf{x+1}) = (x+1)(3x+2)$.

[17] $6x^2 - 2xy - y^2$ is prime because there are no integers m and n such that $mn = -6$ and $m + n = -2$.

[21] $2x^2 + 11x + 5$ can be factored into $(ax+b)(cx+d)$ where $ac = 2$ and $bd = 5$. The correct values for a and c are 2 and 1, and the correct values for b and d are 1 and 5. Note that this gives the correct middle term: $(2x+1)(x+5) = 2x^2 + \mathbf{10x + 1x} + 5 = 2x^2 + \mathbf{11x} + 5$. Therefore $2x^2 + 11x + 5 = (2x+1)(x+5)$.

[25] $5x^2 - 13x + 8 = 5x^2 - 8x - 5x + 8 = x(\mathbf{5x-8}) - 1(\mathbf{5x-8}) = (5x-8)(x-1)$.

[29] $10x^2 - xy + 21y^2$ is prime because there are no integers m and n such that $mn = 210$ and $m + n = -1$.

[33] To factor $x^2 + 10x + 16$, find 2 numbers m and n such that $mn = 16$ and $m + n = 10$. The correct values are 8 and 2. Therefore $x^2 + 10x + 16 = (x+8)(x+2)$.

Exercises 4.5

37. To factor $z^2 + 18z + 32$, find numbers m and n such that $mn = 32$ and $m + n = 18$. Select $m = 16$ and $n = 2$. Therefore $z^2 + 18z + 32 = (z + 16)(z + 2)$.

41. $25x^2 - 22xy + 4y^2$ is prime because there are no integers m and n such that $mn = 100$ and $m + n = -22$.

45. $25x^2 + 25xy + 4y^2 = 25x^2 + 20xy + 5xy + 4y^2 = 5x(5x + 4y) + y(5x + 4y) = (5x + 4y)(5x + y)$.

49. To factor $a^2 - 13ab - 48b^2$ find 2 numbers m and n such that $mn = -48$ and $m + n = -13$. Choose -16 and 3. Then $a^2 - 13ab - 48b^2 = (a - 16b)(a + 3b)$.

53. $6a^2 = 7a + 5$
$6a^2 - 7a - 5 = 0$
$(3a - 5)(2a + 1) = 0$
$3a - 5 = 0$ or $2a + 1 = 0$
$a = \frac{5}{3}$ or $a = -\frac{1}{2}$.

57. $x + 42 = x^2$
$0 = x^2 - x - 42$ or $x^2 - x - 42 = 0$
$(x - 7)(x + 6) = 0$
$x - 7 = 0$ or $x + 6 = 0$
$x = 7$ or $x = -6$.

61. $2t^3 - 8t = 0$
$2t(t^2 - 4) = 0$
$2t(t + 2)(t - 2) = 0$
$2t = 0$ or $t + 2 = 0$ or $t - 2 = 0$
$t = 0$, $t = -2$, or $t = 2$.

65. $18a^2 - 15a - 7 = (3a + 1)(6a - 7)$

69. $8y^2 - 77yz + 16z^2$ is prime because 8×16 has only even factors and no two even numbers can add up to an odd number -77.

73. Multiply $80 \times -3 = -240$ and find numbers m and n such that $mn = -240$ and $m + n = 1$. Select $m = 16$ and $n = -15$. Then $80x^2 + x - 3 = 80x^2 + 16x - 15x - 3 = 16x(5x + 1) - 3(5x + 1) = (16x - 3)(5x + 1)$.

77. $x^3 - 2 + 2x^2 - x = x^3 + 2x^2 - x - 2 = x^2(x + 2) - 1(x + 2) = (x + 2)(x^2 - 1) = (x + 2)(x + 1)(x - 1)$.

[81] $2x^3 - 16 = 2(x^3 - 8) = 2(x^3 - 2^3) = 2(x-2)(x^2 + 2x + 4)$.

[85] $a^2 - b^2 + 6b - 9 = a^2 - (b^2 - 6b + 9) = a^2 - (b-3)^2 = [a+(b-3)][a-(b-3)] = (a+b-3)(a-b+3)$.

[89] $a^2 + b^2 + 2a + 2b + 2ab = (a^2 + 2ab + b^2) + 2a + 2b = (a+b)^2 + 2(a+b) = (a+b)(a+b+2)$.

Review Exercises

[93] Let r be the rate. Then $7r = 28$ which implies that $r = \frac{28}{7} = 4$ mph.

Exercises 4.6

[5] Problem Statement: Find two consecutive odd integers such that their product is 47 more than their sum.

Let x and $x+2$ be the integers. Then $x(x+2) = x + (x+2) + 47$ which reduces to $x^2 + 2x = 2x + 49$ or $x^2 - 49 = 0$. Factoring the left side gives $(x+7)(x-7) = 0$ which implies that $x = -7$ or 7. Therefore the two odd integers can be 7 and 9 or they can be -7 and -5.

[9] Problem Statement: Find three consecutive even integers such that when the product of the 1st two is subtracted from the product of the last two, the result is 32.

(Note that $\Rightarrow$ means implies)

Let x, $x+2$, and $x+4$ be the integers. Then $(x+2)(x+4) - x(x+2) = 32 \Rightarrow x^2 + 6x + 8 - x^2 - 2x = 32 \Rightarrow 4x + 8 = 32 \Rightarrow 4x = 24 \Rightarrow x = \frac{24}{4} = 6$. Therefore the even integers are 6, 8, and 10.

Exercises 4.6

13 Problem Statement: The sum of two numbers is 12 and the sum of their squares is 122. Find the numbers.

Let x and $12-x$ be the numbers. Then
$x^2+(12-x)^2=122 \Rightarrow x^2+144-24x+x^2=122 \Rightarrow 2x^2-24x+22=0 \Rightarrow$
$x^2-12x+11=0$. Factoring the left side gives $(x-11)(x-1)=0 \Rightarrow x=11$ or 1.
If $x=11$ then $12-x=12-11=1$ and if $x=1$ then $12-x=12-1=11$.
In either case, the two numbers are 11 and 1.

21 Problem Statement: The height of a triangle is 4 less than twice the base. The area is 80 in^2. Find the base and height of the triangle.

Let x be the base. Then $2x-4$ is the height and x must satisfy the following equation:
$\frac{1}{2}x(2x-4)=80$ which simplifies to $x^2-2x=80 \Rightarrow x^2-2x-80=0$. Factoring the left side gives $(x-10)(x+8)=0 \Rightarrow x=10$ or $x=-8$.
Therefore the base is 10 inches and the height is $2 \times 10 - 4 = 16$ inches.

25 Let x be the side of the square. Then the perimeter is $4x$ and the area is x^2. Therefore
$x^2=4x+12 \Rightarrow x^2-4x-12=0 \Rightarrow (x-6)(x+2)=0$ and $x=6$.
So the side is 6, the area is 36, and the perimeter is 24.

29 Problem Statement: The length of a rectangle is 3 more than twice its width, and its area is 44 m^2. Find its length and width.

Let x be the width and $2x+3$ the length. Then $x(2x+3)=44 \Rightarrow 2x^2+3x=44 \Rightarrow$
$2x^2+3x-44=0$ which factors to $(2x+11)(x-4)=0$. Therefore $x=4$ and the width is 4 meters and the length is $2 \times 4 + 3 = 11$ meters.

33 Problem Statement: The area of a square is twice the area of a rectangle. The width of the rectangle is 4 less than the side of the square, and the length is 3 more than the side of the square. Find the dimensions of the square and the rectangle.

Let x be the side of the square, $x-4$ the width of the rectangle, and $x+3$ the length of the rectangle. Then $x^2=2(x-4)(x+3) \Rightarrow x^2=2x^2-2x-24 \Rightarrow 0=x^2-2x-24 \Rightarrow$

$0 = (x-6)(x+4) \Rightarrow x = 6$ which means the side of the square is 6, the width of the rectangle is $6 - 4 = 2$, and the length of the rectangle is $6 + 3 = 9$.

Therefore the square is 6 by 6 and the rectangle is 2 by 9 .

[37] **Problem Statement:** Jan went on a 24 mile bicycle ride. Numerically, her time was 2 less than her average rate of travel. Find her rate and time for the trip.

Let x be her average rate for the trip and $x - 2$ her time for the trip. Then

$x(x-2) = 24 \Rightarrow x^2 - 2x = 24 \Rightarrow x^2 - 2x - 24 = 0 \Rightarrow (x-6)(x+4) = 0 \Rightarrow x = 6$.

Therefore her average rate is 6 and the time for the trip is 4 .

[41] **Problem Statement:** James Bond is pushed out of an airplane. Ignoring air resistance, how far would he fall in 10 seconds?

Since $d = 16\, t^2$, $d = 16 \times 10^2 = 1600$ feet.

[45] In example 7, a ball is thrown vertically upward with an initial speed of 32 ft/sec. How long will it take the ball to reach its maximum height of 16 feet.

Since $h = rt - 16\, t^2$ we have $16 = 32t - 16t^2$ or $16t^2 - 32t + 16 = 0$ which simplifies to $t^2 - 2t + 1 = 0$. This factors to $(t - 1)^2 = 0$ which means $t = 1$ second. Therefore the ball will reach its maximum height in 1 second.

[49] **Problem Statement:** Twice Alan's age 12 years from now is equal to the square of his present age. What is his present age?

Let x be his present age. Then $2(x+12) = x^2 \Rightarrow 2x + 24 = x^2 \Rightarrow 0 = x^2 - 2x - 24 \Rightarrow 0 = (x-6)(x+4)$. Hence $x = 6$ and Alan's present age is 6 .

[53] **Problem Statement:** The area of a trapezoid is 30 in^2. The longer base is twice the height, and the shorter base is 3 more than height. Find the height of the trapezoid.

Let h be the height. Then $2h$ is the longer base and $h + 3$ is the shorter base. Since $A = \frac{1}{2}h(b_1 + b_2)$ we have $30 = \frac{1}{2}h[(2h)+(h+3)] \rightarrow 60 = 3h^2 + 3h \Rightarrow 0 = h^2 + h - 20$.

This factors to $0 = (h+5)(h-4) \Rightarrow h = 4$. Therefore the height is 4 inches and the bases are 8 inches and 7 inches.

Exercises 4.6

57 Problem Statement: A movie house contains 2400 seats. The number of rows is 20 less than twice the number of chairs per row. How many rows and chairs per row are in the movie house?

Let x be the number of chairs per row and $2x - 20$ the number of rows. x must satisfy the following equation: $x(2x - 20) = 2400$ or $2x^2 - 20x = 2400$ which simplifies to $x^2 - 10x - 1200 = 0 \Rightarrow (x - 40)(x + 30) = 0 \Rightarrow x = 40$.

Therefore the number of chairs per row is 40 and the number of rows is $2 \times 40 - 20 = 60$ rows.

61 $d = 16t^2 = 16 \times (.8)^2 = 10.24$.

65 $h = 144t - 16t^2 = 144 \times 4.25 - 16 \times (4.25)^2 = 612 - 289 = 323$.

69 $h = 144t - 16t^2 = 144 \times 8.999 - 16 \times (8.999)^2 = 1295.856 - 1295.712 = .144$.

73 Since $D = \frac{n}{2}(n - 3)$ we have $5 = \frac{n}{2}(n - 3)$ which simplifies to $10 = n(n - 3)$ or $0 = n^2 - 3n - 10$. This factors to $0 = (n - 5)(n + 2)$ which implies that $n = 5$ sides.

Chapter Review Exercises

1. From the rules of divisibility we see that 2, 3, 9, and 6 are factors of 54.

5. The prime factorization of 65 is 5×13.

9. $54 = 2 \times 3^3$ and $114 = 2 \times 3 \times 19$. Hence the GCF is $2 \times 3 = 6$.

13. $15xy - 6yz + 14xz$ is prime.

17. $20t^3 - 125t = 5t(4t^2 - 25) = 5t(2t+5)(2t-5)$.   21. $ax - 3x - ay - 3y$ is prime.

25. $x^2 - x - 72 = (x-9)(x+8)$.   29. $20x^2 + 7x - 6 = (4x+3)(5x-2)$.

33. $(x+5)^2 - z^2 = (x+5+z)(x+5-z)$.   37. $x^2 - 36 = 0 \Rightarrow (x+6)(x-6) = 0 \Rightarrow x = \pm 6$.

41. $(s+3)^2 = s^2 + 4s + 19 \Rightarrow s^2 + 6s + 9 = s^2 + 4s + 19 \Rightarrow 6s + 9 = 4s + 19 \Rightarrow 2s = 10 \Rightarrow s = 5$.

45. Let x be the base of the triangle and $2x - 3$ its height. Since $A = \frac{1}{2}bh$ we have
    $\frac{1}{2}x(2x-3) = 10$ $\quad\Leftarrow\quad$ Multiply both sides by 2 to eliminate the fraction
    $x(2x-3) = 20$ $\quad\Leftarrow\quad$ Eliminate the parentheses and move the 20 to the left side
    $2x^2 - 3x - 20 = 0$ $\quad$ or $\quad (2x+5)(x-4) = 0$. Thus x must be 4 (not $-\frac{5}{2}$).
    Therefore the base is 4 meters and the height is $8 - 3$ or 5 meters.

49. Since $h = rt - 16t^2$ we have $32 = 48t - 16t^2$ or $16t^2 - 48t + 32 = 0$. This reduces to $t^2 - 3t + 2 = 0$ or $(t-2)(t-1) = 0$. Then $t = 1$ or 2. Therefore the ball will be 32 feet above the ground at $t = 1$ second and $t = 2$ second. Note at 1 second, it is going up, and at 2 seconds it is going down.

Chapter Test

1. $138 = 2 \times 69 = 2 \times 3 \times 23$. Therefore the prime factorization of 138 is $2 \times 3 \times 23$.

5. $x^3 - 125 = x^3 - 5^3 = (x-5)(x^2 + 5x + 25)$.   9. $a^2 + 6ab - 27b^2 = (a+9b)(a-3b)$.

13. $x^3 - 3x^2 - 25x + 75 = x^2(x-3) - 25(x-3) = (x-3)(x^2 - 25) = (x-3)(x+5)(x-5)$.

17. $2x^2 + 5x = 3$
    $2x^2 + 5x - 3 = 0$
    $(2x-1)(x+3) = 0$
    $x = \frac{1}{2}$ or -3.

21. $18x^3 - 12x^2 + 2x = 0$
    $2x(9x^2 - 6x + 1) = 0$
    $2x(3x-1)^2 = 0$
    $x = 0$ or $\frac{1}{3}$.

25. Let r be the rate and $r-1$ the time. Then $r(r-1) = 3[r + (r-1)] + 3 \Rightarrow$
    $r^2 - r = 3(2r-1) + 3 \Rightarrow r^2 - 7r = 0 \Rightarrow r(r-7) = 0$. Therefore the solutions to the equation are $r = 0$ or 7. But $r = 0$ is no good since $r - 1 = 0 - 1 = -1$. We don't want a negative time. Therefore the rate is 7 mph, the time is 6 hours, and the distance is $7 \times 6 = 42$ miles.

Chapter 5: Rational Expressions

Exercises 5.1

[1] $\frac{6}{x-3}$ is defined for all x where $x - 3 \neq 0$. Therefore the domain is $\{x/x \neq 3\}$.

[5] $\frac{a}{a^2-1}$ is defined for all a where $a^2 - 1 \neq 0$. $a^2 - 1 = 0$ when $a = \pm 1$. Therefore the domain is $\{a/a \neq \pm 1\}$.

[9] To find the domain see where $2x^2 + 11x - 6 = 0$. Factoring we get $(2x - 1)(x + 6) = 0$. Hence $x = \frac{1}{2}$ or -6. Therefore the domain of $\frac{3}{2x^2 + 11x - 6}$ is $\{x/ \neq \frac{1}{2}$ or $-6\}$.

[13] If $x = 10$ then $\frac{2x+4}{3x-2} = \frac{2 \times 10 + 4}{3 \times 10 - 2} = \frac{24}{28} = \frac{6 \times 4}{7 \times 4} = \frac{6}{7}$.

[17] If $a = 1$ and $b = -2$ then $\frac{2a^2 - 3ab - 2b^2}{a^2 - b^2} = \frac{2 \cdot (1)^2 - 3(1)(-2) - 2(-2)^2}{(1)^2 - (-2)^2} = \frac{2 + 6 - 8}{1 - 4} = \frac{0}{-3} = 0$.

[21] If $x = 1.3$ and $y = -2.7$ then $\frac{x^2 - 3xy}{x^2y + xy^2} = \frac{(1.3)^2 - 3(1.3)(-2.7)}{(1.3)^2(-2.7) + (1.3)(-2.7)^2} = \frac{1.69 + 10.53}{-4.563 + 9.477} = \frac{12.22}{4.914} \approx 2.5$.

[25] $\frac{-(3x-2)}{x+4} = \frac{-3x-2}{x+4}$ is false. $\frac{-(3x-2)}{x+4} = \frac{-3x+2}{x+4}$.

[29] $\frac{x^2 + 2x + 1}{x+1} = x^2 + 2$ is false. $\frac{x^2 + 2x + 1}{x+1} = \frac{(x+1)(x+1)}{x+1} = x + 1$.

[33] $\frac{-50}{75} = \frac{-2 \times 25}{3 \times 25} = \frac{-2}{3}$.

[37] $\frac{8-10}{10-8} = \frac{-2}{2} = -1$.

[41] $\frac{30x^8 y^{20}}{15x^{30} y^2} = \frac{30}{15} \cdot \frac{x^8}{x^{30}} \cdot \frac{y^{20}}{y^2} = 2 \cdot \frac{1}{x^{22}} \cdot y^{18} = \frac{2y^{18}}{x^{22}}$.

[45] $\frac{3x+3}{5x+5} = \frac{3(x+1)}{5(x+1)} = \frac{3}{5}$.

[49] $\frac{a^2 - b^2}{a - b} = \frac{(a+b)(a-b)}{a-b} = a + b$.

Exercises 5.1

53 $\dfrac{x^2+3x+2}{x+2} = \dfrac{(x+2)(x+1)}{x+2} = x+1$.

57 $\dfrac{x^2-y^2}{(x+y)^2} = \dfrac{(x+y)(x-y)}{(x+y)(x+y)} = \dfrac{x-y}{x+y}$.

61 $\dfrac{x^2-x-12}{x^2-16} = \dfrac{(x-4)(x+3)}{(x-4)(x+4)} = \dfrac{x+3}{x+4}$.

65 $\dfrac{5y-6-y^2}{3y^2+y-14} = \dfrac{-y^2+5y-6}{(3y+7)(y-2)} = \dfrac{-1\cdot(y^2-5y+6)}{(3y+7)(y-2)} = \dfrac{-1\cdot(y-3)(y-2)}{(3y+7)(y-2)} = -\dfrac{y-3}{3y+7}$.

69 $\dfrac{a^2-49}{ab-ac+7b-7c} = \dfrac{(a-7)(a+7)}{a(b-c)+7(b-c)} = \dfrac{(a-7)(a+7)}{(b-c)(a+7)} = \dfrac{a-7}{b-c}$.

73 $\dfrac{8x+8y}{4x^3+4y^3} = \dfrac{8(x+y)}{4(x^3+y^3)} = \dfrac{8(x+y)}{4(x+y)(x^2-xy+y^2)} = \dfrac{2}{x^2-xy+y^2}$.

77 $\dfrac{2x^3-8x}{x^3-12-4x+3x^2} = \dfrac{2x(x^2-4)}{x^3+3x^2-4x-12} = \dfrac{2x(x^2-4)}{x^2(x+3)-4(x+3)} = \dfrac{2x(x^2-4)}{(x+3)(x^2-4)} = \dfrac{2x}{x+3}$.

81 $\dfrac{x^4-5x^2+4}{x^3+2x^2-x-2} = \dfrac{(x^2-4)(x^2-1)}{x^2(x+2)-1\cdot(x+2)} = \dfrac{(x+2)(x-2)(x^2-1)}{(x^2-1)(x+2)} = x-2$.

Review Problems

89 0 has no reciprocal .

93 $5\times 10 \div 2 = 50 \div 2 = 25$.

Exercises 5.2

1 $\dfrac{4}{7}\cdot\dfrac{2}{5} = \dfrac{4\cdot 2}{7\cdot 5} = \dfrac{8}{35}$.

5 $\dfrac{3}{xy}\cdot\dfrac{5x}{9} = \dfrac{3\cdot 5x}{3\cdot 3xy} = \dfrac{5}{3y}$.

9 $\dfrac{a+2}{a^2-2a+1}\cdot\dfrac{a-1}{a^2-4} = \dfrac{(a+2)(a-1)}{(a-1)(a-1)(a+2)(a-2)} = \dfrac{1}{(a-1)(a-2)}$.

Exercises 5.2

13 $\dfrac{3}{4} \div 9 = \dfrac{3}{4} \cdot \dfrac{1}{9} = \dfrac{3}{4 \cdot 3 \cdot 3} = \dfrac{1}{4 \cdot 3} = \dfrac{1}{12}$

17 $\dfrac{x^2+x}{4y} \div \dfrac{x^2-1}{6y^2} = \dfrac{x(x+1)}{4y} \cdot \dfrac{6y^2}{x^2-1} = \dfrac{x(x+1) \cdot 6y^2}{4y(x+1)(x-1)} = \dfrac{3xy}{2(x-1)}$.

21 $-\dfrac{6}{5} \cdot \dfrac{10}{-9} \div 2 = \dfrac{-6 \cdot 10}{5 \cdot (-9)} \div 2 = \dfrac{-60}{-45} \div 2 = \dfrac{4}{3} \div 2 = \dfrac{4}{3} \cdot \dfrac{1}{2} = \dfrac{4}{6} = \dfrac{2}{3}$.

25 $\dfrac{x^3-1}{10} \div \dfrac{x^2-1}{6} = \dfrac{x^3-1}{10} \cdot \dfrac{6}{x^2-1} = \dfrac{6(x-1)(x^2+x+1)}{10(x+1)(x-1)} = \dfrac{3(x^2+x+1)}{5(x+1)}$.

29 $(x^2-16) \div \dfrac{x^3+5x^2+4x}{x^3+x} = (x+4)(x-4) \cdot \dfrac{x^3+x}{x^3+5x^2+4x} = (x+4)(x-4) \cdot \dfrac{x(x^2+1)}{x(x^2+5x+4)} =$

$\dfrac{(x+4)(x-4)x(x^2+1)}{x(x+4)(x+1)} = \dfrac{(x-4)(x^2+1)}{x+1}$.

33 $\dfrac{ac+bc-ab-b^2}{a^2-b^2} \cdot \dfrac{b-a}{c-b} = \dfrac{c(a+b)-b(a+b)}{(a+b)(a-b)} \cdot \dfrac{-(a-b)}{c-b} = \dfrac{-(a+b)(c-b)(a-b)}{(a+b)(a-b)(c-b)} = -1$.

37 $\dfrac{6u^2-10u-4}{9u^2-1} \cdot \dfrac{3u^2-u}{6u-12} = \dfrac{2(3u^2-5u-2)}{(3u+1)(3u-1)} \cdot \dfrac{u(3u-1)}{6(u-2)} = \dfrac{2u(3u+1)(u-2)(3u-1)}{6(3u+1)(3u-1)(u-2)} = \dfrac{u}{3}$.

41 $\dfrac{y^3+6y^2+9y}{4y^2+13y+3} \div \dfrac{3y+9}{4y^2-11y-3} \cdot \dfrac{81}{y^{18}} = \dfrac{y^3+6y^2+9y}{4y^2+13y+3} \cdot \dfrac{4y^2-11y-3}{3y+9} \cdot \dfrac{81}{y^{18}} =$

$\dfrac{y(y+3)(y+3)}{(4y+1)(y+3)} \cdot \dfrac{(4y+1)(y-3)}{3(y+3)} \cdot \dfrac{81}{y^{18}} = \dfrac{27(y-3)}{y^{17}}$.

45 The distance traveled at $\dfrac{4r}{5}$ mph for $\dfrac{15}{2}$ hours is $\dfrac{4r}{5} \cdot \dfrac{15}{2} = \dfrac{60r}{10} = 6r$ miles.

Review Problems

53 $x^2 - (x-6) = x^2 - x + 6$.

Exercises 5.3

1. $\frac{7}{13} + \frac{11}{13} = \frac{7+11}{13} = \frac{18}{13}.$

5. $\frac{-8}{30} + \frac{-2}{15} = \frac{-4}{15} + \frac{-2}{15} = \frac{-4+(-2)}{15} = \frac{-6}{15} = \frac{-2 \times 3}{5 \times 3} = -\frac{2}{5}.$

9. $\frac{3x}{x^3} + \frac{x+1}{x^2} = \frac{3}{x^2} + \frac{x+1}{x^2} = \frac{3+x+1}{x^2} = \frac{x+4}{x^2}.$

13. $\frac{2}{5} - \frac{1}{5} = \frac{2-1}{5} = \frac{1}{5}.$

17. $\frac{1}{10} - \frac{3}{-10} = \frac{1}{10} + \frac{-3}{-10} = \frac{1}{10} + \frac{3}{10} = \frac{4}{10} = \frac{2}{5}.$

21. $\frac{5}{9z} - \frac{2}{9z} = \frac{3}{9z} = \frac{1}{3z}.$

25. $\frac{7}{10ab} - \frac{-3}{10ab} + \frac{9}{10ab} = \frac{7-(-3)+9}{10ab} = \frac{7+3+9}{10ab} = \frac{19}{10ab}.$

29. $\frac{8y(9)}{8y} = 9$ is true because we can cancel the factor of $8y$ in the numerator and denominator.

33. $\frac{2}{x} - \frac{x-4}{x} = \frac{-x-2}{x}$ is false; $\frac{2}{x} - \frac{x-4}{x} = \frac{2-(x-4)}{x} = \frac{2-x+4}{x} = \frac{6-x}{x}.$

37. $\frac{x}{x^2+9} + \frac{3}{x^2+9} = \frac{x+3}{x^2+9}.$

41. $\frac{3x}{x^2} + \frac{7}{x} = \frac{3}{x} + \frac{7}{x} = \frac{10}{x}.$

45. $\frac{7}{x-y} + \frac{1}{x-y} - \frac{3}{y-x} = \frac{7}{x-y} + \frac{1}{x-y} - \frac{-1 \cdot 3}{-1 \cdot (y-x)} = \frac{7}{x-y} + \frac{1}{x-y} - \frac{-3}{x-y} = \frac{11}{x-y}.$

49. $\frac{15x}{1-5x} - \frac{4}{5x-1} = \frac{-15x}{-(1-5x)} - \frac{4}{5x-1} = \frac{-15x}{5x-1} - \frac{4}{5x-1} = \frac{-15x-4}{5x-1}$ or $\frac{-(15x+4)}{5x-1}.$

53. $\frac{x^4}{x^4+x^2} - \frac{1}{x^4+x^2} = \frac{x^4-1}{x^4+x^2} = \frac{(x^2+1)(x^2-1)}{x^2(x^2+1)} = \frac{x^2-1}{x^2}.$

Exercises 5.3

[57] $\dfrac{x^2+1}{x^4-81} + \dfrac{6x+8}{x^4-81} = \dfrac{x^2+1+6x+8}{x^4-81} = \dfrac{x^2+6x+9}{x^4-81} = \dfrac{(x+3)(x+3)}{(x^2+9)(x^2-9)} =$

$\dfrac{(x+3)(x+3)}{(x^2+9)(x+3)(x-3)} = \dfrac{x+3}{(x^2+9)(x-3)}$.

[61] To find the LCD of $\dfrac{5}{6}$ and $\dfrac{1}{9}$, note that $6 = 2 \times 3$ and $9 = 3^2$. The LCD $= 2^a \cdot 3^b$ where a is the highest power of 2 and b is the highest power of 3. Therefore $a = 1$ and $b = 2$ and the LCD is $2^1 \cdot 3^2 = 2 \cdot 9 = 18$.

[65] To find the LCD of $\dfrac{4}{9}, \dfrac{1}{8},$ and $\dfrac{1}{6}$ note that $9 = 3^2$, $8 = 2^3$, and $6 = 2 \times 3$. Therefore the LCD is $2^3 \cdot 3^2 = 8 \cdot 9 = 72$.

[69] $8x^3 = 2^3 x^3$ and $2x = 2x$. Therefore the LCD of $\dfrac{1}{8x^3}$ and $\dfrac{3}{2x}$ is $2^3 x^3$ or $8x^3$.

[73] $12x^3 = 2^2 \cdot 3x^3$ and $48x^5 = 2^4 \cdot 3x^5$. Therefore the LCD of $\dfrac{1}{12x^3}$ and $\dfrac{7}{48x^5}$ is $2^4 \cdot 3x^5 = 48x^5$.

[77] $8k = 2^3 k$ and $4k^2 - 4k = 4k(k-1) = 2^2 k(k-1)$. Hence the LCD of $\dfrac{3}{8k}$ and $\dfrac{1}{4k^2-4k}$ is $2^3 \cdot k \cdot (k-1) = 8k(k-1)$.

[81] To find the LCD of $\dfrac{1}{a-4}$ and $\dfrac{2}{4-a}$, note that $\dfrac{2}{4-a} = \dfrac{-1 \cdot 2}{-1 \cdot (4-a)} = \dfrac{-2}{a-4}$.
Therefore the LCD is $a-4$.

[85] $x^3 - y^3 = (x-y)(x^2+xy+y^2)$ and $x^2 - y^2 = (x+y)(x-y)$. Therefore the LCD of $\dfrac{x}{x^3-y^3}$ and $\dfrac{y}{x^2-y^2}$ is $(x+y)(x-y)(x^2+xy+y^2)$.

[89] $5a^2 - 17ab + 6b^2 = (5a-2b)(a-3b)$ and $6a^2 - 17ab - 3b^2 = (6a+b)(a-3b)$.
The LCD is $(5a-2b)(a-3b)(6a+b)$.

[93] The LCD of $\dfrac{1}{3x}, \dfrac{1}{3x+1},$ and $\dfrac{1}{x+1}$ is $3x(3x+1)(x+1)$.

Exercises 5.3

97 $x^2 + 2x + xy + 2y = (x+2)(x+y)$ and $x^2 + 2x - xy - 2y = (x+2)(x-y)$.
Therefore the LCD is $(x+2)(x+y)(x-y)$.

Review Problems

101 $3 + 5 \times 3 = 3 + 15 = 18$.

Exercises 5.4

1 Yes; $\frac{21}{28} = \frac{3 \times 7}{4 \times 7} = \frac{3}{4}$.

5 Yes; $\frac{x-3}{7-x} = \frac{-(x-3)}{-(7-x)} = \frac{3-x}{x-7}$.

9 No; $\frac{3}{2x+2}$ is not equivalent to $\frac{3}{x+2}$.

13 True; $\frac{10}{25} + \frac{6}{5} = \frac{2}{5} + \frac{6}{5} = \frac{8}{5}$.

17 True; $\frac{6ab^2 - 4b^2}{2b^2} = \frac{2b^2(3a-2)}{2b^2} = 3a - 2$.

21 $\frac{11}{6} + \frac{9}{4} = \frac{11}{6} \cdot \frac{2}{2} + \frac{9}{4} \cdot \frac{3}{3} = \frac{22}{12} + \frac{27}{12} = \frac{49}{12}$.

25 $\frac{2}{a^2} + \frac{3}{4} = \frac{2}{a^2} \cdot \frac{4}{4} + \frac{3}{4} \cdot \frac{a^2}{a^2} = \frac{8}{4a^2} + \frac{3a^2}{4a^2} = \frac{3a^2 + 8}{4a^2}$.

29 Since $12x^{16} = 2^2 \cdot 3 x^{16}$ and $18x^3 = 2 \cdot 3^2 x^3$, the LCD $= 2^a \cdot 3^b \cdot x^c$ where a is the highest power of 2, b is the highest power of 3, and c is the highest power of x. Therefore the LCD $= 2^2 \cdot 3^2 \cdot x^{16} = 36 x^{16}$. Then

$$\frac{1}{12x^{16}} + \frac{5}{18x^3} = \frac{1}{12x^{16}} \cdot \frac{3}{3} + \frac{5}{18x^3} \cdot \frac{2x^{13}}{2x^{13}} = \frac{3}{36x^{16}} + \frac{10x^{13}}{36x^{16}} = \frac{10x^3 + 3}{36x^{16}}.$$

33 $\frac{8}{10} - \frac{-6}{-5} = \frac{4}{5} - \frac{6}{5} = -\frac{2}{5}$.

37 $\frac{a}{3a-3} - \frac{a+1}{a-a^2} = \frac{a}{3(a-1)} - \frac{a+1}{a(1-a)} = \frac{a}{3(a-1)} - \frac{-(a+1)}{-a(1-a)} = \frac{a}{3(a-1)} + \frac{a+1}{a(a-1)} =$

$\frac{a}{3(a-1)} \cdot \frac{a}{a} + \frac{a+1}{a(a-1)} \cdot \frac{3}{3} = \frac{a^2}{3a(a-1)} + \frac{3a+3}{3a(a-1)} = \frac{a^2 + 3a + 3}{3a(a-1)}$.

Exercises 5.4

41 $\frac{4}{7} + \frac{3}{7} \cdot \frac{2}{3} = \frac{4}{7} + \frac{3 \times 2}{7 \times 3} = \frac{4}{7} + \frac{2}{7} = \frac{6}{7}$.

45 $\left(\frac{1}{4} + \frac{5}{6}\right) \div \left(\frac{4}{9} - \frac{1}{6}\right) = \left(\frac{3}{12} + \frac{10}{12}\right) \div \left(\frac{8}{18} - \frac{3}{18}\right) = \frac{13}{12} \div \frac{5}{18} = \frac{13}{12} \times \frac{18}{5} = \frac{13 \times 18}{12 \times 5} = \frac{13 \times 6 \times 3}{2 \times 6 \times 5} = \frac{39}{10}$.

49 $\frac{8}{5y} - \frac{2}{5y} - \frac{6}{5y} = \frac{8-2-6}{5y} = \frac{0}{5y} = 0$.

53 $\frac{3}{y^2+3y} - \frac{y}{3y+9} = \frac{3}{y(y+3)} - \frac{y}{3(y+3)} = \frac{3}{y(y+3)} \cdot \frac{3}{3} - \frac{y}{3(y+3)} \cdot \frac{y}{y} =$

$\frac{9}{3y(y+3)} - \frac{y^2}{3y(y+3)} = \frac{9-y^2}{3y(y+3)} = \frac{(3+y)(3-y)}{3y(y+3)} = \frac{3-y}{3y}$.

57 $x+5-\frac{x-2}{x+3} = \frac{x+5}{1} \cdot \frac{x+3}{x+3} - \frac{x-2}{x+3} = \frac{x^2+8x+15}{x+3} - \frac{x-2}{x+3} = \frac{x^2+8x+15-(x-2)}{x+3} =$

$\frac{x^2+8x+15-x+2}{x+3} = \frac{x^2+7x+17}{x+3}$.

61 $\frac{x^2-1}{3x^2-3x} + \frac{2x+8}{6x} = \frac{(x+1)(x-1)}{3x(x-1)} + \frac{2(x+4)}{6x} = \frac{x+1}{3x} + \frac{x+4}{3x} = \frac{x+1+x+4}{3x} = \frac{2x+5}{3x}$.

65 $\frac{x}{2x^2-5x-3} + \frac{4}{2x^2+7x+3} = \frac{x}{(2x+1)(x-3)} + \frac{4}{(2x+1)(x+3)} =$

$\frac{x}{(2x+1)(x-3)} \cdot \frac{x+3}{x+3} + \frac{4}{(2x+1)(x+3)} \cdot \frac{x-3}{x-3} =$

$\frac{x(x+3)}{(2x+1)(x-3)(x+3)} + \frac{4(x-3)}{(2x+1)(x+3)(x-3)} = \frac{x^2+3x+4x-12}{(2x+1)(x-3)(x+3)} =$

$\frac{x^2+7x-12}{(2x+1)(x-3)(x+3)}$.

69 $\left(\frac{x^4}{2y^2}\right)^3 + \left(\frac{y^7}{3x^3}\right)^2 = \frac{x^{12}}{8y^6} + \frac{y^{14}}{9x^6} = \frac{x^{12}}{8y^6} \cdot \frac{9x^6}{9x^6} + \frac{y^{14}}{9x^6} \cdot \frac{8y^6}{8y^6} = \frac{9x^{18}}{72x^6y^6} + \frac{8y^{20}}{72x^6y^6} = \frac{9x^{18}+8y^{20}}{72x^6y^6}$.

Exercises 5.4

73 $\dfrac{2}{x+3}+\dfrac{3}{3-x}-\dfrac{x-x^2}{x^2-9}=\dfrac{2}{x+3}+\dfrac{-3}{x-3}-\dfrac{x-x^2}{(x+3)(x-3)}=$

$\dfrac{2}{x+3}\cdot\dfrac{x-3}{x-3}+\dfrac{-3}{x-3}\cdot\dfrac{x+3}{x+3}-\dfrac{x-x^2}{(x+3)(x-3)}=\dfrac{2(x-3)-3(x+3)-(x-x^2)}{(x+3)(x-3)}=$

$\dfrac{2x-6-3x-9-x+x^2}{(x+3)(x-3)}=\dfrac{x^2-2x-15}{(x+3)(x-3)}=\dfrac{(x+3)(x-5)}{(x+3)(x-3)}=\dfrac{x-5}{x-3}.$

77 $\dfrac{1}{x}-\dfrac{2x-6}{x^2-6x}+\dfrac{1}{x-6}=\dfrac{1}{x}-\dfrac{2x-6}{x(x-6)}+\dfrac{1}{x-6}=\dfrac{1}{x}\cdot\dfrac{x-6}{x-6}-\dfrac{2x-6}{x(x-6)}+\dfrac{1}{x-6}\cdot\dfrac{x}{x}=$

$\dfrac{x-6-(2x-6)+x}{x(x-6)}=\dfrac{x-6-2x+6+x}{x(x-6)}=\dfrac{0}{x(x-6)}=0.$

81 Let x be the number. Then the sum of $\tfrac{1}{4}$ of a number, $\tfrac{1}{3}$ of a number, and $\tfrac{1}{2}$ of a number is

$\tfrac{1}{4}x+\tfrac{1}{3}x+\tfrac{1}{2}x=\dfrac{x}{4}+\dfrac{x}{3}+\dfrac{x}{2}=\dfrac{x}{4}\cdot\dfrac{3}{3}+\dfrac{x}{3}\cdot\dfrac{4}{4}+\dfrac{x}{2}\cdot\dfrac{6}{6}=\dfrac{3x+4x+6x}{12}=\dfrac{13x}{12}.$

85 The perimeter is $\dfrac{x}{x+1}+\dfrac{3}{x-1}+\dfrac{2}{x^2-1}=\dfrac{x}{x+1}\cdot\dfrac{x-1}{x-1}+\dfrac{3}{x-1}\cdot\dfrac{x+1}{x+1}+\dfrac{2}{(x+1)(x-1)}=$

$\dfrac{x(x-1)+3(x+1)+2}{(x+1)(x-1)}=\dfrac{x^2-x+3x+3+2}{(x+1)(x-1)}=\dfrac{x^2+2x+5}{(x+1)(x-1)}.$

Review Problems

89 $\dfrac{11}{3}\div\dfrac{6}{x}=\dfrac{11}{3}\cdot\dfrac{x}{6}=\dfrac{11x}{18}.$

Exercises 5.5

1 $\dfrac{\frac{4}{3}}{\frac{7}{5}}=\dfrac{4}{3}\div\dfrac{7}{5}=\dfrac{4}{3}\times\dfrac{5}{7}=\dfrac{20}{21}.$

5 $\dfrac{1}{\frac{a}{b}}=1\div\dfrac{a}{b}=1\times\dfrac{b}{a}=\dfrac{b}{a}.$

Exercises 5.5

9 $\dfrac{\frac{3a-3}{b^3}}{\frac{a^2-a}{b^9}} = \dfrac{3a-3}{b^3} \div \dfrac{a^2-a}{b^9} = \dfrac{3(a-1)}{b^3} \times \dfrac{b^9}{a(a-1)} = \dfrac{3b^6}{a}$.

13 $\dfrac{2}{\frac{1}{a}+\frac{1}{b}} = \dfrac{2}{\frac{b}{ab}+\frac{a}{ba}} = \dfrac{2}{\frac{b+a}{ab}} = 2 \div \dfrac{b+a}{ab} = 2 \times \dfrac{ab}{a+b} = \dfrac{2ab}{a+b}$.

17 $\dfrac{\frac{m+n}{m}}{\frac{1}{m}+\frac{1}{n}} = \dfrac{\frac{m+n}{m}}{\frac{n}{mn}+\frac{m}{nm}} = \dfrac{\frac{m+n}{m}}{\frac{m+n}{mn}} = \dfrac{m+n}{m} \div \dfrac{m+n}{mn} = \dfrac{m+n}{m} \times \dfrac{mn}{m+n} = n$.

21 $\dfrac{2+\frac{3}{x^2}}{x-\frac{1}{5}} = \dfrac{\left(2+\frac{3}{x^2}\right)\cdot 5x^2}{\left(x-\frac{1}{5}\right)\cdot 5x^2} = \dfrac{10x^2+15}{5x^3-x^2}$.

25 $\dfrac{\frac{3}{x}-\frac{3}{x+h}}{h} = \dfrac{\left(\frac{3}{x}-\frac{3}{x+h}\right)\cdot x(x+h)}{h\cdot x(x+h)} = \dfrac{\frac{3}{x}\cdot x(x+h)-\frac{3}{x+h}\cdot x(x+h)}{h\cdot x(x+h)} = \dfrac{3(x+h)-3x}{h\,x(x+h)} =$

$\dfrac{3x+3h-3x}{h\,x(x+h)} = \dfrac{3h}{h\,x(x+h)} = \dfrac{3}{x(x+h)}$.

29 $\dfrac{\frac{1}{4}}{2x-1} = \dfrac{\frac{1}{4}\cdot 4}{(2x-1)\cdot 4} = \dfrac{1}{(2x-1)\cdot 4} = \dfrac{1}{8x-4}$.   **33** $\dfrac{\frac{a^2b}{6x}}{\frac{a^7b}{3x^2}} = \dfrac{\frac{a^2b}{6x}\cdot 6x^2}{\frac{a^7b}{3x^2}\cdot 6x^2} = \dfrac{xa^2b}{2a^7b} = \dfrac{x}{2a^5}$.

37 $\dfrac{\frac{s}{r-s}}{\frac{r+s}{s}+\frac{s}{r-s}} = \dfrac{\frac{s}{r-s}\cdot s(r-s)}{\left(\frac{r+s}{s}+\frac{s}{r-s}\right)\cdot s(r-s)} = \dfrac{s^2}{\frac{r+s}{s}\cdot s(r-s)+\frac{s}{r-s}\cdot s(r-s)} =$

$\dfrac{s^2}{(r+s)(r-s)+s^2} = \dfrac{s^2}{r^2-s^2+s^2} = \dfrac{s^2}{r^2}$.

41 $\dfrac{\dfrac{(2x^4)^3}{x^2-14x+48}}{\dfrac{(4x^8)^2}{8x^2-64x}} = \dfrac{\dfrac{8x^{12}}{(x-8)(x-6)}}{\dfrac{16x^{16}}{8x(x-8)}} = \dfrac{\dfrac{8x^{12}}{(x-8)(x-6)}}{\dfrac{16x^{16}}{8x(x-8)}} \cdot \dfrac{8x(x-8)(x-6)}{8x(x-8)(x-6)} =$

$\dfrac{8x^{12} \cdot 8x}{16x^{16} \cdot (x-6)} = \dfrac{64x^{13}}{16x^{16}(x-6)} = \dfrac{4}{x^3(x-6)}.$

45 $2 + \dfrac{2}{2+\dfrac{2}{x}} = 2 + \dfrac{2 \cdot x}{\left(2+\dfrac{2}{x}\right) \cdot x} = 2 + \dfrac{2x}{2x+2} = 2 + \dfrac{\cancel{2}x}{\cancel{2}(x+1)} = 2 \cdot \dfrac{x+1}{x+1} + \dfrac{x}{x+1} =$

$\dfrac{2x+2}{x+1} + \dfrac{x}{x+1} = \dfrac{3x+2}{x+1}.$

49 $t + \dfrac{\dfrac{1}{9t^2}-1}{\dfrac{1}{3}-\dfrac{1}{9t}} = t + \dfrac{\dfrac{1}{9t^2}-1}{\dfrac{1}{3}-\dfrac{1}{9t}} \cdot \dfrac{9t^2}{9t^2} = t + \dfrac{1-9t^2}{3t^2-t} = t + \dfrac{(1+3t)(1-3t)}{t(3t-1)} =$

$t + \dfrac{(1+3t)(-1)(3t-1)}{t(3t-1)} = t \cdot \dfrac{t}{t} + \dfrac{-(1+3t)}{t} = \dfrac{t^2-1-3t}{t} = \dfrac{t^2-3t-1}{t}.$

53 $1 + \dfrac{1}{1+\dfrac{1}{1+\dfrac{1}{2}}} = 1 + \dfrac{1}{1+\dfrac{1 \cdot 2}{\left(1+\dfrac{1}{2}\right) \cdot 2}} = 1 + \dfrac{1}{1+\dfrac{2}{2+1}} = 1 + \dfrac{1}{1+\dfrac{2}{3}} = 1 + \dfrac{1}{\dfrac{3}{3}+\dfrac{2}{3}} =$

$1 + \dfrac{1}{\dfrac{5}{3}} = 1 + \dfrac{3}{5} = \dfrac{5}{5} + \dfrac{3}{5} = \dfrac{8}{5}.$

Exercises 5.5

Review Problems

[57] The domain of $\frac{7}{x+1}$ is $\{x/x \neq -1\}$.

[61] $x^2 + 3x - 10 = 0$ factors to $(x+5)(x-2) = 0$ which means $x = -5$ or 2.

Exercises 5.6

[5] To solve $\frac{y-2}{6} + \frac{y}{2} = 5$ multiply both sides by the LCD 6:

$\left(\frac{y-2}{6} + \frac{y}{2}\right) \cdot 6 = 5 \cdot 6 \Rightarrow \frac{y-2}{6} \cdot 6 + \frac{y}{2} \cdot 6 = 30 \Rightarrow y - 2 + 3y = 30 \Rightarrow$

$4y = 32 \Rightarrow y = 8$.

[9] To solve $\frac{x}{x+1} = \frac{9}{11}$ multiply both sides by the LCD $11(x+1)$:

$\frac{x}{x+1} \cdot 11(x+1) = \frac{9}{11} \cdot 11(x+1) \Rightarrow 11x = 9(x+1) \Rightarrow 11x = 9x + 9 \Rightarrow$

$2x = 9 \Rightarrow x = \frac{9}{2}$.

[17] To solve $\frac{2}{a-2} + 2 = \frac{a}{a-2}$ multiply both sides by $a - 2$:

$\left(\frac{2}{a-2} + 2\right) \cdot (a-2) = \frac{a}{a-2} \cdot (a-2) \Rightarrow \frac{2}{a-2} \cdot (a-2) + 2 \cdot (a-2) = a \Rightarrow$

$2 + 2a - 4 = a \Rightarrow a = 2$.

But $a = 2$ is not in the domain of the equation. Therefore there are no solutions.

[21] To solve $\frac{-10x}{x^2+1} = 3$ multiply both sides by $x^2 + 1$:

$\frac{-10x}{x^2+1} \cdot (x^2+1) = 3 \cdot (x^2+1) \Rightarrow -10x = 3x^2 + 3 \Rightarrow 0 = 3x^2 + 10x + 3 \Rightarrow$

$0 = (3x+1)(x+3) \Rightarrow x = -\frac{1}{3}$ or $x = -3$.

Exercises 5.6

25 To solve $\frac{x+4}{5} = \frac{3}{10}$ multiply both sides by 10 :

$\frac{x+4}{5} \cdot 10 = \frac{3}{10} \cdot 10 \;\Rightarrow\; 2(x+4) = 3 \;\Rightarrow\; 2x+8 = 3 \;\Rightarrow\; x = -\frac{5}{2}$.

29 $\frac{x+3}{2} = \frac{x+2}{2} + \frac{1}{x} \;\Rightarrow\; 2x \cdot \left(\frac{x+3}{2}\right) = 2x \cdot \left(\frac{x+2}{2} + \frac{1}{x}\right) \;\Rightarrow\;$

$\frac{2x \cdot (x+3)}{2} = \frac{2x \cdot (x+2)}{2} + \frac{2x \cdot 1}{x} \;\Rightarrow\; x(x+3) = x(x+2) + 2 \;\Rightarrow\;$

$x^2 + 3x = x^2 + 2x + 2 \;\Rightarrow\; x = 2$.

33 $\frac{1}{x-2} = \frac{9-x^2}{x^2+x-6}$ is equivalent to $\frac{1}{x-2} = \frac{(3+x)(3-x)}{(x+3)(x-2)}$ or $\frac{1}{x-2} = \frac{3-x}{x-2}$.

Multiply both sides by $x-2$:

$(x-2) \cdot \frac{1}{x-2} = (x-2) \cdot \frac{3-x}{x-2} \;\Rightarrow\; 1 = 3-x \;\Rightarrow\; x = 2$.

But $x = 2$ is not in the domain of the equation. Therefore there are no solutions.

37 $\frac{3}{x+1} - \frac{x}{x-1} = \frac{\frac{3}{x^2} - 1}{1 - \frac{1}{x^2}}$ implies that $\frac{3}{x+1} - \frac{x}{x-1} = \frac{\left(\frac{3}{x^2} - 1\right) \cdot x^2}{\left(1 - \frac{1}{x^2}\right) \cdot x^2}$ or

$\frac{3}{x+1} - \frac{x}{x-1} = \frac{3-x^2}{x^2-1}$. Multiply both sides by $(x+1)(x-1)$:

$\left(\frac{3}{x+1} - \frac{x}{x-1}\right) \cdot (x+1)(x-1) = \frac{3-x^2}{(x+1)(x-1)} \cdot (x+1)(x-1)$ which simplifies to

$3(x-1) - x(x+1) = 3 - x^2 \;\Rightarrow\; 3x - 3 - x^2 - x = 3 - x^2 \;\Rightarrow\; 2x = 6 \;\Rightarrow\; x = 3$.

41 $\frac{4}{p-3} - \frac{4p^2}{p^3} = \frac{1}{15}$ simplifies to $\frac{4}{p-3} - \frac{4}{p} = \frac{1}{15}$. Then multiply both sides by $15p(p-3)$:

$15p(p-3) \cdot \left(\frac{4}{p-3} - \frac{4}{p}\right) = 15p(p-3) \cdot \frac{1}{15} \;\Rightarrow\; 60p - 60(p-3) = p(p-3) \;\Rightarrow\;$

$60p - 60p + 180 = p^2 - 3p \Rightarrow 0 = p^2 - 3p - 180 \Rightarrow (p-15)(p+12) = 0$.

Therefore $p = 15$ or -12.

$\boxed{45}$ $\dfrac{3a-5}{a^2-3a+2} - \dfrac{2a}{1-a} = \dfrac{a+5}{a-2}$ is equivalent to $\dfrac{3a-5}{(a-2)(a-1)} - \dfrac{-2a}{-(1-a)} = \dfrac{a+5}{a-2} \Rightarrow$

$\dfrac{3a-5}{(a-2)(a-1)} + \dfrac{2a}{a-1} = \dfrac{a+5}{a-2}$. Multiply both sides by $(a-1)(a-2)$:

$(a-1)(a-2) \cdot \left[\dfrac{3a-5}{(a-2)(a-1)} + \dfrac{2a}{a-1} \right] = (a-1)(a-2) \cdot \dfrac{a+5}{a-2} \Rightarrow$

$3a - 5 + 2a(a-2) = (a+5)(a-1) \Rightarrow 3a - 5 + 2a^2 - 4a = a^2 + 4a - 5 \Rightarrow$

$a^2 - 5a = 0 \Rightarrow a(a-5) = 0 \Rightarrow a = 0$ or 5.

$\boxed{49}$ $\dfrac{2x^3+9}{x} = \dfrac{8x+9}{2} \Rightarrow 2x \cdot \dfrac{2x^3+9}{x} = 2x \cdot \dfrac{8x+9}{2} \Rightarrow 4x^3 + 18 = 8x^2 + 9x \Rightarrow$

$4x^3 - 8x^2 - 9x + 18 = 0$. The right side can be factored by grouping:

$4x^2(x-2) - 9(x-2) = 0 \Rightarrow (x-2)(4x^2 - 9) = 0 \Rightarrow (x-2)(2x+3)(2x-3) = 0$.

Therefore $x = 2, -\dfrac{3}{2},$ or $\dfrac{3}{2}$.

$\boxed{53}$ $P = 2L + 2W \Rightarrow P - 2L = 2W \Rightarrow W = \dfrac{P-2L}{2}$.

$\boxed{57}$ To solve $\dfrac{3x}{4} - \dfrac{2y}{3} = 2$ for y we can 1st eliminate the fractions by multiplying both sides by 12:

$12 \cdot \dfrac{3x}{4} - 12 \cdot \dfrac{2y}{3} = 12 \cdot 2 \Rightarrow 9x - 8y = 24 \Rightarrow 9x - 24 = 8y \Rightarrow y = \dfrac{9x-24}{8}$.

$\boxed{61}$ To solve $\dfrac{1}{f} = \dfrac{1}{d} + \dfrac{1}{D}$ for f multiply both sides by fdD to eliminate the fractions:

$fdD \cdot \dfrac{1}{f} = fdD \cdot \dfrac{1}{d} + fdD \cdot \dfrac{1}{D} \Rightarrow dD = fD + fd$. Now factor out the f on the right side:

$dD = f(D+d)$. Finally divide both sides by $D+d$ to isolate f:

$\dfrac{dD}{D+d} = \dfrac{f(D+d)}{D+d}$ which implies that $f = \dfrac{dD}{D+d}$.

Exercises 5.6

65 To solve $y = \dfrac{2x+1}{3x-7}$ for x multiply both sides by $3x-7$ to eliminate the fractions:

$y(3x-7) = \dfrac{2x+1}{3x-7} \cdot (3x-7) \;\Rightarrow\; 3xy - 7y = 2x + 1$.

Get all the terms with x on the left side of the equation and factor out the x:

$3xy - 2x = 7y + 1 \;\Rightarrow\; x(3y-2) = 7y+1 \;\Rightarrow\; \dfrac{x(3y-2)}{3y-2} = \dfrac{7y+1}{3y-2} \;\Rightarrow\; x = \dfrac{7y+1}{3y-2}$.

69 $w = \dfrac{1+x}{1-x} \;\Rightarrow\; w(1-x) = 1+x \;\Rightarrow\; w - wx = 1 + x$.

Get all the terms with x on the right side and factor out the x:

$w - 1 = x + wx \;\Rightarrow\; w - 1 = x(1+w) \;\Rightarrow\; x = \dfrac{w-1}{w+1}$.

73 $T = \dfrac{w_2 m_1}{m_1 + m_2} \;\Rightarrow\; (m_1 + m_2)T = w_2 m_1 \;\Rightarrow\; m_1 T + m_2 T = w_2 m_1$.

Isolate the term with m_2: $m_2 T = w_2 m_1 - m_1 T$. Finally divide both sides by T:

$m_2 = \dfrac{w_2 m_1 - m_1 T}{T}$.

77 $Q = \dfrac{w(H-h)}{T-t} \;\Rightarrow\; (T-t)Q = w(H-h) \;\Rightarrow\; TQ - tQ = wH - wh$.

Isolate the term with H: $TQ - tQ + wh = wH$.

Finally divide both sides by w: $H = \dfrac{TQ - tQ + wh}{w}$.

Review Problems

85 Let x, $x+1$, and $x+2$ be the numbers. Then $x + x + 1 + x + 2 = 39 \;\Rightarrow\; 3x + 3 = 39 \;\Rightarrow\; 3x = 36 \;\Rightarrow\; x = 12$. Therefore the numbers are 12, 13, and 14.

Exercises 5.7

[5] Problem Statement: The sum of a number and twice its reciprocal is $7\frac{2}{7}$. Find the number.

Let x be the number. Then $x + 2 \cdot \frac{1}{x} = 7\frac{2}{7}$. Since $7\frac{2}{7} = \frac{49}{7} + \frac{2}{7} = \frac{51}{7}$ the equation can be rewritten as $x + \frac{2}{x} = \frac{51}{7}$. Multiply both sides of the equation by $7x$:

$7x \cdot \left(x + \frac{2}{x}\right) = 7x \cdot \frac{51}{7} \Rightarrow 7x^2 + 14 = 51x \Rightarrow 7x^2 - 51x + 14 = 0 \Rightarrow$
$(7x - 2)(x - 7) = 0 \Rightarrow x = 7 \text{ or } \frac{2}{7}$.

[9] Problem Statement: The numerator of a fraction is 3 more than the denominator. If the denominator is multiplied by 15 and 10 is added to the numerator, then the new fraction is $\frac{1}{2}$. Find the original fraction.

Let x be the denominator and $x + 3$ the numerator. Then $\frac{x+3+10}{15x} = \frac{1}{2} \Rightarrow \frac{x+13}{15x} = \frac{1}{2} \Rightarrow$
$2 \cdot (x + 13) = 15x \Rightarrow 2x + 26 = 15x \Rightarrow 26 = 13x \Rightarrow x = 2$.
Therefore the fraction is $\frac{5}{2}$.

[13] Problem Statement: A number is added to the numerator and the denominator of $\frac{5}{7}$. If 1 is added to this new fraction, the result is $\frac{9}{5}$. What is the number?

Let x be the number. Then $\frac{5+x}{7+x} + 1 = \frac{9}{5}$. Multiply both sides by $5(7 + x)$:

$5(7+x) \cdot \left[\frac{5+x}{7+x} + 1\right] = 5(7+x) \cdot \frac{9}{5} \Rightarrow 5(5+x) + 5(7+x) = 9(7+x) \Rightarrow$
$25 + 5x + 35 + 5x = 63 + 9x \Rightarrow 10x + 60 = 63 + 9x \Rightarrow x = 3$.

[17] Problem Statement: When 30 is divided by a number, both the quotient and the remainder are equal to 3. Find the number.

Let x be the number. Then $\frac{30}{x} = 3 + \frac{3}{x}$. Multiply both sides by x which gives
$30 = 3x + 3 \Rightarrow 27 = 3x \Rightarrow x = 9$. Therefore the number is 9.

[21] Problem Statement: One car travels 200 km in the same time that a second car travels 150 km. If the first car travels 30 km/h faster than the second car, find the speed of each car.

Exercises 5.7

Let x be the speed of the 2nd car and $x+30$ the speed of the 1st car. Since $T = \frac{D}{R}$, the time for the 1st car is $\frac{200}{x+30}$ and the time for the 2nd car is $\frac{150}{x}$.
Since the two times are the same,

$$\frac{200}{x+30} = \frac{150}{x} \Rightarrow 200x = 150(x+30) \Rightarrow 200x = 150x + 4500 \Rightarrow 50x = 4500 \Rightarrow$$

$x = \frac{4500}{50} = 90$.

Hence the speed of the 2nd car is 90 km/h and the speed of the 1st car is 120 km/h.

25 Problem Statement: Sara can jog twice as fast as she can walk. She goes for a 5 hour walk and then jogs for 2 hours. If she jogs 3 miles less than she walks, how far does she walk?

Let x be her speed walking and $2x$ her speed jogging. Since $D = RT$, she travels $5x$ on her 5 hour walk and she travels $2(2x) = 4x$ on her 2 hour jog. Therefore $5x - 3 = 4x \Rightarrow x = 3$.
Therefore Sara walks $5 \cdot 3 = 15$ miles.

29 Problem Statement: The speed of a small plane in still air is 120 mph. In one hour and fifteen minutes the plane flies 70 miles with the wind and 75 miles against the wind. Find the speed of the wind and the time required for each part of the trip.

Let x be the speed of the wind. Then the plane has a speed of $120 + x$ with the wind and a speed of $120 - x$ against the wind. Since $T = \frac{D}{R}$, the plane travels $\frac{70}{120+x}$ hours with the wind and $\frac{75}{120-x}$ hours against the wind. The total time is one hour and fifteen minutes $= 1 + \frac{15}{60} = 1 + \frac{1}{4} = \frac{5}{4}$. Therefore $\frac{70}{120+x} + \frac{75}{120-x} = \frac{5}{4}$. Multiply both sides by $4(120+x)(120-x)$:

$$4(120+x)(120-x) \cdot \left[\frac{70}{120+x} + \frac{75}{120-x}\right] = 4(120+x)(120-x) \cdot \frac{5}{4} \Rightarrow$$

$280(120-x) + 300(120+x) = 5(120+x)(120-x) \Rightarrow$

$33{,}600 - 280x + 36{,}000 + 300x = 72{,}000 - 5x^2 \Rightarrow 5x^2 + 20x - 2400 = 0 \Rightarrow$

$x^2 + 4x - 480 = 0 \Rightarrow (x-20)(x+24) = 0 \Rightarrow x = 20$.

Therefore the speed of the wind is 20 mph, the time traveling with the wind is $\frac{70}{120+20} = \frac{70}{140} = \frac{1}{2}$ hr, and the the time traveling against the wind is $\frac{75}{120-20} = \frac{75}{100} = \frac{3}{4}$ hr.

33 If $C = \frac{600}{x+40}$ and $x = 60$ then $C = \frac{600}{60+40} = \frac{600}{100} = \$6$.

Exercises 5.7

[37] If $C = \dfrac{600}{x+40}$ and $C = 5$ then $5 = \dfrac{600}{x+40}$ $\Rightarrow$ $5(x+40) = 600$ $\Rightarrow$ $5x + 200 = 600$ $\Rightarrow$ $5x = 400$ $\Rightarrow$ $x = \dfrac{400}{5} = 80$.

[41] Problem Statement: Two resistors are connected in parallel, one of then with a resistance of 24 ohms. If the total resistance is 6 ohms, find the value of the other resistor.

Use the formula $\dfrac{1}{R} = \dfrac{1}{R_1} + \dfrac{1}{R_2}$ where $R = 6$ and $R_1 = 24$. R_2 is the value of the other resistor. Therefore $\dfrac{1}{6} = \dfrac{1}{24} + \dfrac{1}{R_2}$. Multiply both sides of the equation by $24R_2$ which gives $4R_2 = R_2 + 24$ $\Rightarrow$ $3R_2 = 24$ $\Rightarrow$ $R_2 = 8$ ohms.

[45] Since $F = \dfrac{9}{5}C + 32$, $98.4 = \dfrac{9}{5}C + 32$. Then $98.4 - 32 = \dfrac{9}{5}C$. To solve for C, multiply both sides of the equation by $\dfrac{5}{9}$: $\dfrac{5}{9} \cdot (98.4 - 32) = \dfrac{5}{9} \cdot \dfrac{9}{5}C$ $\Rightarrow$ $C = \dfrac{5}{9} \cdot 66.4 = 36.9°$.

[49] Problem Statement: Gary drove 400 miles in 9 hours. His rate during the last 200 miles was 10 mph faster than his rate during the first 200 miles. Find his rate during the first 200 miles of the trip.

Let r be the rate for the 1st 200 miles and $r + 10$ the rate for the last 200 miles. Since $T = \dfrac{D}{R}$ and the total time is 9 hours, $\dfrac{200}{r} + \dfrac{200}{r+10} = 9$. Multiply both sides by $r(r+10)$:

$r(r+10) \cdot \left[\dfrac{200}{r} + \dfrac{200}{r+10} \right] = r(r+10) \cdot 9$ $\Rightarrow$ $200(r+10) + 200r = 9r^2 + 90r$ $\Rightarrow$

$200r + 2000 + 200r = 9r^2 + 90r$ $\Rightarrow$ $0 = 9r^2 - 310r - 2000$ $\Rightarrow$ $(r-40)(9r+50) = 0$ $\Rightarrow$

$r = 40$. The rate for the 1st 200 miles is 40 mph.

Review Problems

[53] $\dfrac{x}{2} = \dfrac{7}{12}$ $\Rightarrow$ $12x = 2 \cdot 7$ $\Rightarrow$ $12x = 14$ $\Rightarrow$ $x = \dfrac{14}{12} = \dfrac{7}{6}$.

[57] $A = \dfrac{1}{2} \cdot h \cdot b = \dfrac{1}{2} \cdot 8 \cdot 2 = 8$ cm^2.

Exercises 5.8

1 $6 : 24 = \frac{6}{24} = \frac{1}{4} = 1 : 4$.

5 $20 : 12 : 8 = 5(4) : 3(4) : 2(4) = 5 : 3 : 2$.

9 The ratio of 40 feet to 60 feet is $\frac{40 \text{ ft}}{60 \text{ ft}} = \frac{2 \times 20}{3 \times 20} = \frac{2}{3}$.

13 The ratio of 4 days to 3 weeks is $\frac{4 \text{ days}}{3 \text{ weeks}} = \frac{4 \text{ days}}{21 \text{ days}} = \frac{4}{21}$.

17 Problem Statement: Three numbers are in the ratio $2 : 3 : 4$. Their sum is 63. Find the numbers.

The 3 numbers can be represented as $2x$, $3x$, and $4x$. Since their sum is 63,

$2x + 3x + 4x = 63 \Rightarrow 9x = 63 \Rightarrow x = 7$.

Therefore the 3 numbers are 2×7, 3×7, and 4×7 or 14, 21, and 28 .

21 Problem Statement: The ratio of the height to the base of a triangle is $3 : 5$. If its area is 120 in^2, find its height and base.

Let $3x$ be the height and $5x$ the base. Since $A = \frac{1}{2}bh$, $\frac{1}{2}(5x)(3x) = 120 \Rightarrow \frac{15x^2}{2} = 120$. Multiply both sides by 2: $15x^2 = 240$ or $x^2 = 16 \Rightarrow x = 4$.

Hence the height is $3 \times 4 = 12$ inches and the base is $5 \times 4 = 20$ inches.

25 Problem Statement: Rey has \$5.20 worth of dimes and quarters. The ratio of the number of dimes to the number of quarters is $3 : 4$. Find the number of dimes and quarters.

Let $3x$ be the number of dimes and $4x$ the number of quarters. The value of the dimes is $.10(3x)$ and the value of the quarters is $.25(4x)$. Since the total value is 5.20,

$.10(3x) + .25(4x) = 5.20 \Rightarrow .3x + 1.0x = 5.20 \Rightarrow 1.3x = 5.20 \Rightarrow x = \frac{5.20}{1.3} = 4$.

Therefore the number of dimes is $3 \times 4 = 12$ and the number of quarters is $4 \times 4 = 16$.

29 Problem Statement: Avi drove 450 miles. The ratio of his rate (in mph) to his time (in hours) was $9 : 2$. Find his rate and time.

Let $9x$ be his rate and $2x$ his time. Since $D = RT$,

$(9x)(2x) = 450 \Rightarrow 18x^2 = 450 \Rightarrow x^2 = 25 \Rightarrow x = 5$.

Therefore his rate is 45 mph and his time is 10 hours.

$\boxed{33}$ $\dfrac{6}{x} = \dfrac{11}{x+11} \Rightarrow 6(x+11) = 11x \Rightarrow 6x + 66 = 11x \Rightarrow 6 = 5x \Rightarrow x = \dfrac{6}{5}$.

$\boxed{37}$ $\dfrac{3x-4}{-3} = \dfrac{8-6x}{6} \Rightarrow 6(3x-4) = -3(8-6x) \Rightarrow 18x - 24 = -24 + 18x$.

Since both sides are the same, the equation is satisfied by all real numbers.

$\boxed{41}$ $\dfrac{a}{b} = \dfrac{c}{x}$ implies that $ax = bc$ or $a = \dfrac{bc}{x}$.

$\boxed{45}$ $\dfrac{P_i V_i}{T_i} = \dfrac{P_f V_f}{T_f}$ implies that $P_i V_i T_f = P_f V_f T_i$. To solve for T_i divide both sides of the equation by $P_f V_f$: $\dfrac{P_i V_i T_f}{P_f V_f} = \dfrac{P_f V_f T_i}{P_f V_f} \Rightarrow T_i = \dfrac{P_i V_i T_f}{P_f V_f}$.

$\boxed{49}$ $\dfrac{b-a}{c-b} = \dfrac{a}{b} \Rightarrow b(b-a) = a(c-b) \Rightarrow b^2 - ab = ac - ab$. Cancel the $-ab$ on both sides giving $b^2 = ac$. Therefore $a = \dfrac{b^2}{c}$.

$\boxed{53}$ $\dfrac{14}{21} = \dfrac{15}{10}$ is false because 14×10 or 140 is not equal to 21×15 or 315.

$\boxed{57}$ $\dfrac{x^2 - 16}{x+4} = \dfrac{x-4}{1}$ is true because $(x^2 - 16) \cdot 1 = x^2 - 16 = (x+4)(x-4)$.

$\boxed{61}$ Since 5 dozen cookies requires 3 cups of flour, $\dfrac{5 \text{ dozen cookies}}{3 \text{ cups of flour}} = \dfrac{8 \text{ dozen cookies}}{x \text{ cups of flour}}$. This can also be expressed as $\dfrac{3 \text{ cups of flour}}{5 \text{ dozen cookies}} = \dfrac{x \text{ cups of flour}}{8 \text{ dozen cookies}}$. Therefore the equation $\dfrac{x}{8} = \dfrac{3}{5}$ is true.

$\boxed{65}$ Since $\dfrac{5}{3} = \dfrac{8}{x}$ this implies that $5x = 3 \cdot 8$. Dividing both sides by $3 \cdot 5$ gives $\dfrac{\not{5}x}{3 \cdot \not{5}} = \dfrac{\not{3} \cdot 8}{\not{3} \cdot 5}$ or $\dfrac{x}{3} = \dfrac{8}{5}$. Therefore the equation is true.

Exercises 5.8

[69] Problem Statement: A recipe for 3 cups of cherry sauce requires 3/4 of a cup of orange juice. How many cups of orange juice are needed for 8 cups of sauce?

Let x be the number of cups of orange juice needed. Therefore

$$\frac{3 \text{ cups of cherry sauce}}{3/4 \text{ of a cup of orange juice}} = \frac{8 \text{ cups of cherry sauce}}{x \text{ cups of orange juice}} \text{ or more simply } \frac{3}{3/4} = \frac{8}{x} \text{ which implies}$$

that $3x = \frac{3}{4} \cdot 8 \;\Rightarrow\; 3x = 6$. Therefore $x = \frac{6}{3} = 2$ cups of orange juice.

[73] Problem Statement: A chain saw requires 2 cans of gas to cut 3/8 of a cord of wood. How many cans of gas are needed to cut $1\frac{1}{2}$ cords of wood?

Let x be the number of cans of gas needed. Then $\dfrac{2 \text{ cans of gas}}{3/8 \text{ of cord of wood}} = \dfrac{x \text{ cans of gas}}{1\frac{1}{2} \text{ cords of wood}}$ or more simply $\dfrac{2}{3/8} = \dfrac{x}{3/2}$ where $1\frac{1}{2}$ is replaced by $\frac{3}{2}$. Now cross multiply:

$2 \cdot \frac{3}{2} = x \cdot \frac{3}{8} \;\Rightarrow\; 3 = \frac{3x}{8} \;\Rightarrow\; 24 = 3x \;\Rightarrow\; x = 8$ cans of gas.

[77] Problem Statement: Andy has to write a 2000 word report. After writing 2 pages he found that they contain 325 words. At this rate how many pages must he write to complete the paper?

Let x be the total number of pages. Then $\dfrac{2 \text{ pages}}{325 \text{ words}} = \dfrac{x \text{ pages}}{2000 \text{ words}} \;\Rightarrow\; \dfrac{2}{325} = \dfrac{x}{2000} \;\Rightarrow\;$

$4000 = 325x \;\Rightarrow\; x = \dfrac{4000}{325} = 12.3$. Therefore he must write 10.3 pages to complete the paper.

[81] Problem Statement: If 2.2 pounds = 1 kilogram, how many kilograms are in 53 pounds?

Let x be the number of kilograms. Then $\dfrac{2.2 \text{ pounds}}{1 \text{ kilogram}} = \dfrac{53 \text{ pounds}}{x \text{ kilograms}} \;\Rightarrow\; 2.2x = 1(53) \;\Rightarrow\;$ $x = \dfrac{53}{2.2} = 24.09$ kilograms.

[85] Since $\dfrac{V_1}{P_2} = \dfrac{V_2}{P_1}$ and $V_1 = 2.0$ liters, $P_1 = 5.0$ atm, and $V_2 = 25$ liters we have the following

equation: $\dfrac{2.0}{P_2} = \dfrac{25}{5.0} \;\Rightarrow\; 2(5) = 25 P_2 \;\Rightarrow\; P_2 = \dfrac{10}{25} = \dfrac{2}{5}$ atm.

Exercises 5.8

89 To solve $\frac{b-a}{c-b} = \frac{c-a}{c+b}$ for a first cross multiply to eliminate the fractions:

$(b-a)(c+b) = (c-b)(c-a) \Rightarrow bc + b^2 - ac - ab = c^2 - ac - bc + ab$.

Eliminate the $-ac$ on both sides and combine like terms. This gives $2bc + b^2 - c^2 = 2ab$.

To solve for a divide both sides by $2b$. Therefore $a = \frac{2bc + b^2 - c^2}{2b}$.

93 $\frac{V_1}{T_1} = \frac{V_2}{T_2}$ where T_1 and T_2 are expressed in Kelvin degrees. Since $V_1 = 200$ mL,

$T_1 = 0°C = 273°K$, and $T_2 = 27°C = 27 + 273 = 300°K$ we have the following equation:

$\frac{200}{273} = \frac{V_2}{300} \Rightarrow 200 \cdot 300 = 273 V_2 \Rightarrow V_2 = \frac{200 \cdot 300}{273} = 219.78$.

Therefore the gas occupies 219.78 mL.

Chapter Review Exercises

1. Since $\frac{x}{x^2 - x - 12} = \frac{x}{(x-4)(x+3)}$ its domain is $\{x/x \neq -3 \text{ or } 4\}$.

5. $\frac{20x^8}{12x^4} = \frac{5 \cdot 4 \cdot x^{8-4}}{3 \cdot 4} = \frac{5x^4}{3}$.

9. $\frac{10r^8}{9s^4} \cdot \frac{6s^{20}}{25r^{12}} = \frac{(\cancel{2} \cdot 2) \cdot (\cancel{3} \cdot 2)}{(\cancel{3} \cdot 3) \cdot (\cancel{5} \cdot 5)} \cdot \frac{r^8}{r^{12}} \cdot \frac{s^{20}}{s^4} = \frac{4}{15} \cdot \frac{1}{r^4} \cdot \frac{s^{16}}{1} = \frac{4s^{16}}{15r^4}$.

13. $\frac{x+4}{x^2} + \frac{x^2-4}{x^2} - \frac{x+4+x^2-4}{x^2} = \frac{x^2+x}{x^2} = \frac{x(x+1)}{x^2} = \frac{x+1}{x}$.

17. Since $26x^6y^2 = 2 \cdot 13x^6y^2$ and $39x^3y^7 = 3 \cdot 13x^3y^7$ the LCD of $\frac{x+1}{26x^6y^2}$ and $\frac{4}{39x^3y^7}$ is

$2 \cdot 3 \cdot 13 \cdot x^6 \cdot y^7 = 78x^6y^7$.

Chapter 5 Review Exercises and Test

21. $\dfrac{k}{k^2-4} + \dfrac{k-1}{4k-2k^2} = \dfrac{k}{(k+2)(k-2)} + \dfrac{k-1}{2k(2-k)} = \dfrac{k}{(k+2)(k-2)} + \dfrac{-(k-1)}{2k(k-2)} =$

$\dfrac{k \cdot 2k - (k-1)(k+2)}{2k(k+2)(k-2)} = \dfrac{2k^2 - (k^2+k-2)}{2k(k+2)(k-2)} = \dfrac{2k^2 - k^2 - k + 2}{2k(k+2)(k-2)} = \dfrac{k^2 - k + 2}{2k(k+2)(k-2)}.$

25. $\dfrac{\tfrac{2}{9}}{1\tfrac{4}{3}} = \dfrac{\tfrac{2}{9}}{\tfrac{7}{3}} = \dfrac{2}{9} \div \dfrac{7}{3} = \dfrac{2}{9} \times \dfrac{3}{7} = \dfrac{2 \times 3}{3 \times 3 \times 7} = \dfrac{2}{21}.$

29. $\dfrac{2-x}{x} = \dfrac{4}{3} \;\Rightarrow\; 6 - 3x = 4x \;\Rightarrow\; 6 = 7x \;\Rightarrow\; x = \dfrac{6}{7}.$

33. $\dfrac{x}{3} - \dfrac{y}{4} = 1 \;\Rightarrow\; 12\left(\dfrac{x}{3} - \dfrac{y}{4}\right) = 12 \cdot 1 \;\Rightarrow\; 4x - 3y = 12 \;\Rightarrow\; 4x - 12 = 3y \;\Rightarrow\; y = \dfrac{4x-12}{3}.$

37. Let x be the 1st number and $2x$ the 2nd number. Then $\dfrac{1}{x} + \dfrac{1}{2x} = -\dfrac{1}{4} \;\Rightarrow$

$4x \cdot \left(\dfrac{1}{x} + \dfrac{1}{2x}\right) = 4x \cdot \left(-\dfrac{1}{4}\right) \;\Rightarrow\; 4 + 2 = -x \;\Rightarrow\; 6 = -x \;\Rightarrow\; x = -6.$

Therefore the numbers are -6 and -12.

41. Let $2x$ be the time and $3x$ the rate. Then $(2x)(3x) = 24 \;\Rightarrow\; 6x^2 = 24 \;\Rightarrow\; x^2 = 4 \;\Rightarrow\; x = 2$. Therefore the time is 4 hours and the rate is 6 mph.

Chapter Test

1. The domain of $\dfrac{2}{3x-4}$ is $\{x / x \neq \tfrac{4}{3}\}$.

5. $\dfrac{18x^{20}}{32y^6} \cdot \dfrac{24y^{18}}{27x^5} = \dfrac{(\cancel{9} \cdot \cancel{2})(\cancel{8} \cdot \cancel{3})x^{20}y^{18}}{(\cancel{8} \cdot \cancel{2} \cdot 2)(\cancel{9} \cdot \cancel{3})y^6 x^5} = \dfrac{x^{15}y^{12}}{2}.$

9. $\dfrac{x}{x^2-9} - \dfrac{4}{3x-x^2} = \dfrac{x}{(x+3)(x-3)} - \dfrac{4}{x(3-x)} = \dfrac{x}{(x+3)(x-3)} - \dfrac{-4}{x(x-3)} =$

$\dfrac{x}{(x+3)(x-3)} + \dfrac{4}{x(x-3)} = \dfrac{x}{(x+3)(x-3)} \cdot \dfrac{x}{x} + \dfrac{4}{x(x-3)} \cdot \dfrac{x+3}{x+3} = \dfrac{x^2 + 4x + 12}{x(x-3)(x+3)}.$

13. $\dfrac{\frac{a}{b}-\frac{b}{a}}{\frac{a}{b}+\frac{b}{a}-2} = \dfrac{\left(\frac{a}{b}-\frac{b}{a}\right)\cdot ab}{\left(\frac{a}{b}+\frac{b}{a}-2\right)\cdot ab} = \dfrac{a^2-b^2}{a^2+b^2-2ab} = \dfrac{(a+b)(a-b)}{(a-b)(a-b)} = \dfrac{a+b}{a-b}$.

17. $\dfrac{2}{z+1}+\dfrac{1}{z-1} = -\dfrac{2}{3} \Rightarrow 3(z+1)(z-1)\cdot\left[\dfrac{2}{z+1}+\dfrac{1}{z-1}\right] = 3(z+1)(z-1)\cdot\left(-\dfrac{2}{3}\right) \Rightarrow$

$6(z-1)+3(z+1) = -2(z+1)(z-1) \Rightarrow 6z-6+3z+3 = -2z^2+2$.

Then combine like terms putting all terms on the left:

$2z^2+9z-5 = 0 \Rightarrow (2z-1)(z+5) = 0 \Rightarrow z = \dfrac{1}{2}$ or -5.

21. Let x be the number. Then $\dfrac{4}{11+x}+2 = \dfrac{11}{5}$. Multiply both sides by $5(11+x)$:

$5(11+x)\cdot\left[\dfrac{4}{11+x}+2\right] = 5(11+x)\cdot\dfrac{11}{5} \Rightarrow 20+10(11+x) = 11(11+x) \Rightarrow$

$20+110+10x = 121+11x \Rightarrow 130+10x = 121+11x \Rightarrow x = 9$.

25. Let x be the weight of the woman on mars. Then $\dfrac{5}{2} = \dfrac{122}{x} \Rightarrow 5x = 122\cdot 2 \Rightarrow$

$x = \dfrac{244}{5} = 48.8$ pounds.

Chapter 6: Linear Equations in Two Variables

Exercises 6.1

[1]
```
 -8  -6  -4  -2   0   1
         •
```
The point coordinate of the point A is -6.

[5] The graph of the points -5, -3, 1, and 4 is
```
 -6  -5  -4  -3  -2  -1   0   1   2   3   4
     •       •               •           •
```

[9] The graph of $x < 3$ is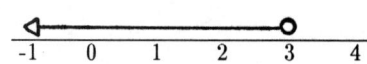

Use this diagram for problems 13, 17, and 21:

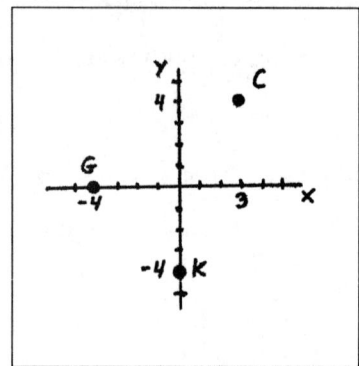

[13] Point C is (3, 4).   [17] Point G is $(-4, 0)$.   [21] Point K is $(0, -4)$

[25] The graph of the points $(-2, -2)$, $(-1, 0)$, $(0, 2)$, and $(1, 4)$ is given below.
A relationship between the x-coordinate and y-coordinate is that the y-coordinate is half the x-coordinate minus 1   or $y = \frac{1}{2}x - 1$.

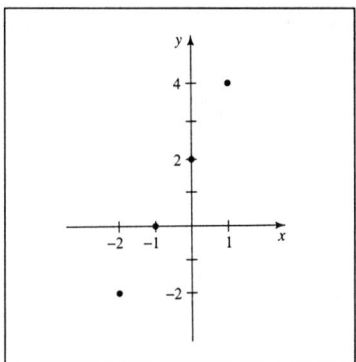

Figure 25

[29] The graph of the points (1, 1), (1, 3), (4, 1), and (4, 3) is given below.

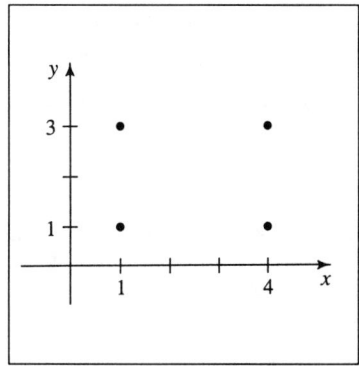

[33] The point (8, 0) is not in any of the 4 quadrants. It lies on the x-axis.

[37] To see if (4, 9) is a solution to the equation $y = 2x + 1$, substitute $x = 4$ and $y = 9$. Then we see that (4, 9) is a solution because $9 = 2 \times 4 + 1$ is a true statement.
Note that $2 \times 4 + 1 = 8 + 1 = 9$.

[41] The ordered pair $(-2, -12)$ is a solution to $y = 6x$ because $-12 = 6 \cdot (-2)$ is a true statement.

[45] The ordered pair (3, 3) is not a solution to $y = -3$ because $3 \neq -3$.

Exercises 6.1

49 If $y = -2x$ then the y-coordinate of $(0, \square)$ is determined by substituting $x = 0$ into the equation. Then $y = -2x = -2 \times 0 = 0$ and the solution is $(0, 0)$.

The x-coordinate of $(\square, 8)$ is determined by substituting $y = 8$ in the equation $y = -2x$ which gives $8 = -2x$. Therefore $x = \frac{8}{-2} = -4$.

Similarly the x-coordinate of $(\square, -6)$ is found by substituting $y = -6$ which gives $-6 = -2x$ and therefore $x = \frac{-6}{-2} = 3$.

So the solutions are $(0, 0), (-4, 8),$ and $(3, -6)$.

53 If $3x + 4y = 7$ then the y-coordinate of $(1, \square)$ is found by making $x = 1$. Then $3 \cdot 1 + 4y = 7 \Rightarrow 3 + 4y = 7 \Rightarrow 4y = 4 \Rightarrow y = 1$. To find the y-coordinate of $(2, \square)$, substitute $x = 2$. Then $3 \cdot 2 + 4y = 7 \Rightarrow 6 + 4y = 7 \Rightarrow 4y = 1 \Rightarrow y = \frac{1}{4}$.

To find the x-coordinate of $(\square, -2)$ make $y = -2$. This means $3x + 4 \cdot (-2) = 7 \Rightarrow 3x - 8 = 7 \Rightarrow 3x = 15 \Rightarrow x = 5$. So the solutions are $(1, 1), (2, \frac{1}{4}),$ and $(5, -2)$.

57 If $y + 7 = 0$ then $y = -7$ regardless of the value of x. Therefore the solutions are $(2, -7), (3, -7),$ and $(70, -7)$.

61 The following is the rectangle with vertices $(1, 1), (3, 1), (-1, -2),$ and $(3, -2)$:

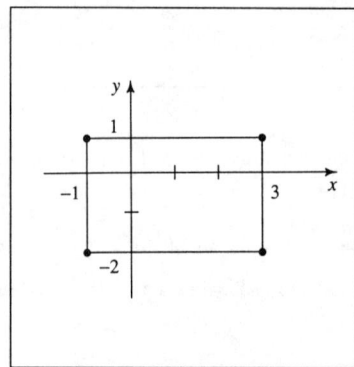

65 The graph below is the line through $(6, 2)$ and $(-3, -4)$. The point C where the line crosses the x-axis is $(3, 0)$ and the point D where the line crosses the y-axis is $(0, -2)$.

Exercises 6.1

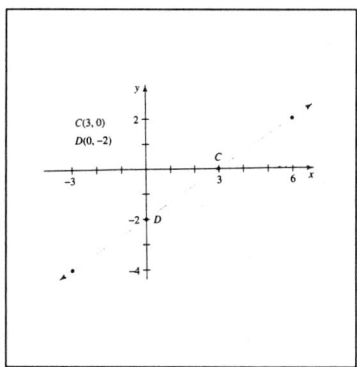

Figure 65

69. The line determined by the points (3, 4) and (3, −2) is a vertical line. The x-coordinate of the point where the y-coordinate is 1 is 3. Note that every point on the line has x-coordinate 3.

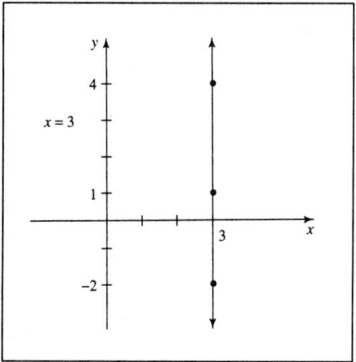

73. 5 points which have the form $(x, 6 - x)$ are (5, 1), (4, 2), (3, 3), (2, 4), and (0, 6).

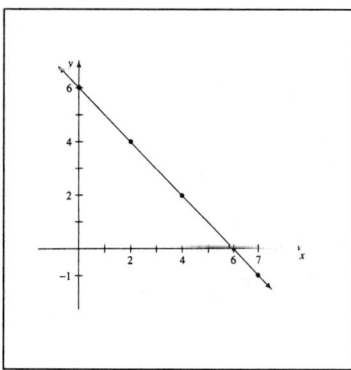

Exercises 6.1

Review Problems

[77] If $2x + 3y = 6$ then $2x = 6 - 3y$ or $x = \dfrac{6-3y}{2}$.

[81] If $\dfrac{2x}{3} - \dfrac{y}{4} = -2$ then $\dfrac{2x}{3} = \dfrac{y}{4} - 2$. Multiply both sides by $\dfrac{3}{2}$ to isolate x:

$\dfrac{3}{2} \cdot \dfrac{2x}{3} = \dfrac{3}{2} \cdot \left(\dfrac{y}{4} - 2\right) \Rightarrow x = \dfrac{3}{2} \cdot \dfrac{y}{4} - \dfrac{3 \cdot 2}{2} \Rightarrow x = \dfrac{3y}{8} - 3$.

Exercises 6.2

[1] To graph $y = 2x + 7$ using a TI-81 graphing calculator, press $\boxed{ZOOM}$ $\boxed{6}$ to set the calculator to the standard graphing screen. Press the $\boxed{Y=}$ key and make $y_1 = 2x + 7$ by pressing the following keys: $\boxed{2}$ $\boxed{X|T}$ $\boxed{+}$ $\boxed{7}$ $\boxed{ENTER}$; and finally press $\boxed{GRAPH}$.

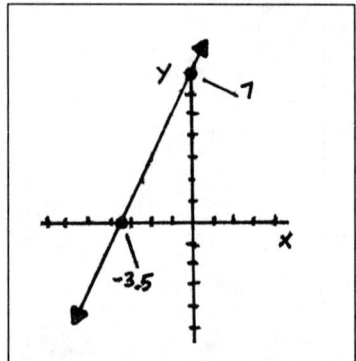

Figure 1

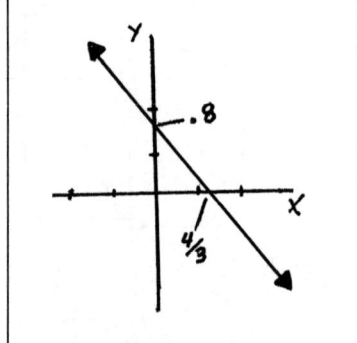

Figure 5

[5] To graph $3x + 5y = 4$ using a graphing calculator, you must first solve for y:

$3x + 5y = 4 \Rightarrow 5y = 4 - 3x \Rightarrow y = \dfrac{4-3x}{5} \Rightarrow y = \dfrac{4}{5} - \dfrac{3}{5}x \Rightarrow y = .8 - .6x$.

Then press the following keys: $\boxed{ZOOM}$ $\boxed{6}$ to set the calculator to the standard graphing screen. Then press $\boxed{Y=}$ $\boxed{.}$ $\boxed{8}$ $\boxed{-}$ $\boxed{.}$ $\boxed{6}$ $\boxed{X|T}$ $\boxed{ENTER}$ $\boxed{GRAPH}$.

Exercises 6.2

[9] $y = \frac{2}{7}x - 2 \Rightarrow 7 \cdot y = 7 \cdot \left(\frac{2}{7}x - 2\right) \Rightarrow 7y = 2x - 14$. Move the $7y$ to the right side and the -14 to the left side. This gives $14 = 2x - 7y$ which is equivalent to $2x - 7y = 14$.

[13] To find the x-intercept of $y = -2x$ set $y = 0$ and solve for x.
Thus $0 = -2x \Rightarrow x = \frac{0}{-2} = 0$. Therefore the x-intercept is 0.
To find the y-intercept set $x = 0$. Then $y = -2 \cdot 0 = 0$. Therefore the y-intercept is 0.
To find a 2nd point on the graph, let $x = 1$. Then $y = -2 \cdot 1 = -2$. Therefore $(1, -2)$ is a point on the graph. Let $x = -1$. Then $y = -2 \cdot (-1) = 2$ and $(-1, 2)$ is a 3rd point on the graph.

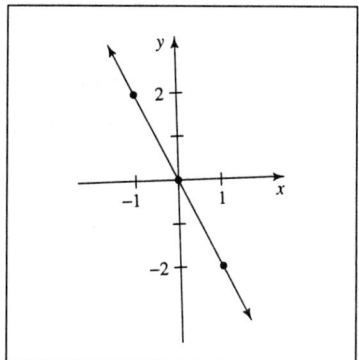

Figure 13

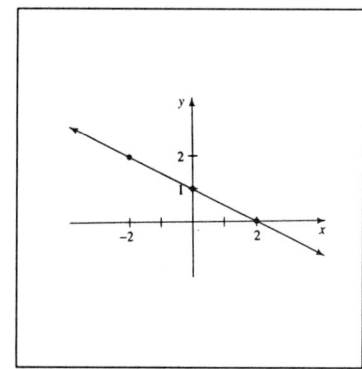

Figure 17

[17] To find the x-intercept of $y = -\frac{1}{2}x + 1$ set $y = 0$. This means $0 = -\frac{1}{2}x + 1 \Rightarrow \frac{1}{2}x = 1 \Rightarrow x = 2$. Therefore the x-intercept is 2.
To find the y-intercept set $x = 0$. Then $y = -\frac{1}{2} \cdot 0 + 1 = 0 + 1 = 1$.
Therefore the y-intercept is 1. To find a 3rd point on the graph let $x = -2$. Then $y = -\frac{1}{2} \cdot (-2) + 1 = 1 + 1 = 2$. So a 3rd point on the line is $(-2, 2)$.

[21] Given $3x - 2y = 6$, if $y = 0$ then $3x - 0 = 6$ or $x = 2$. Therefore the x-intercept is 2.
If $x = 0$ then $0 - 2y = 6 \Rightarrow y = \frac{6}{-2} = -3$. Therefore the y-intercept is -3.

To find a 3rd point, let $x = 4$. Then $12 - 2y = 6 \Rightarrow -2y = -6 \Rightarrow y = \frac{-6}{-2} = 3$. So a 3rd point on the line is $(4, 3)$.

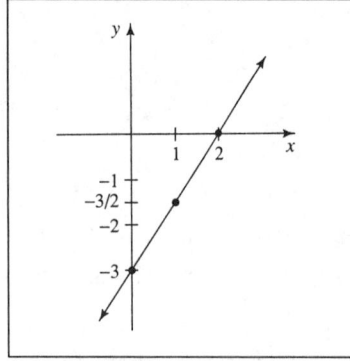

Figure 21

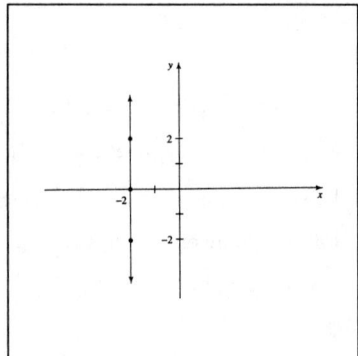

Figure 25

[25] The line given by $x = -2$ is a vertical line. Every point on this line has x-coordinate -2. Therefore the x-intercept is -2 and there is no y-intercept. Two other points on the line are $(-2, 1)$ and $(-2, 2)$.

[29] $x - 8 = 0$ is equivalent to $x = 8$. This line is a vertical line with x-intercept 8. There is no y-intercept. $(8, 1)$ and $(8, 2)$ are two other points on the line.

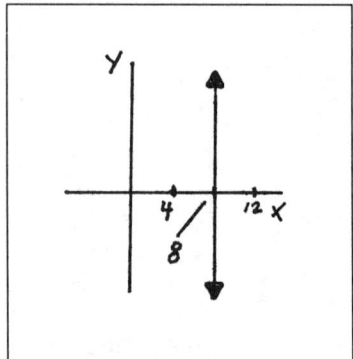

Figure 29

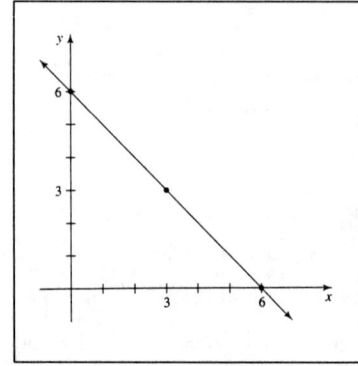

Figure 33

[33] To graph the line $x + y = 6$, let $x = 1$. Then $1 + y = 6$ or $y = 5$. Therefore $(1, 5)$ is a point on the line. If $x = 0$ then $0 + y = 6 \Rightarrow y = 6$. So $(0, 6)$ is a point on the line. Finally, let $x = 6$. Then $6 + y = 6 \Rightarrow y = 0$ and $(6, 0)$ is a 3rd point on the line.

Exercises 6.2

[37] Given $x - 4y = 2$ and if $x = 0$ then $-4y = 2$ or $y = \frac{2}{-4} = -\frac{1}{2}$. So $(0, -\frac{1}{2})$ lies on the line. If $y = 0$ then $x - 0 = 2 \Rightarrow x = 2$ so $(2, 0)$ lies on the line. To get a 3rd point, let $y = -1$. Then $x - 4(-1) = 2 \Rightarrow x + 4 = 2 \Rightarrow x = -2$ and $(-2, -1)$ lies on the line.

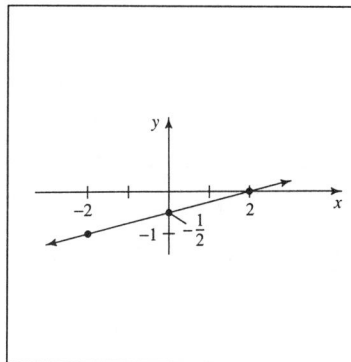

Figure 37

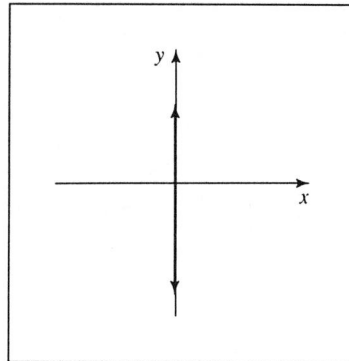

Figure 41

[41] $x = 0$ is the vertical line consisting of all the points whose x-coordinate is 0. This line is the y-axis. Three points on this line are $(0, 0)$, $(0, 1)$, and $(0, 2)$.

[45] $2y = 3$ is equivalent to $y = \frac{3}{2}$. This is the horizontal line consisting of all the points whose y-coordinate is $\frac{3}{2}$. Three points on this line are $(0, \frac{3}{2})$, $(1, \frac{3}{2})$, and $(2, \frac{3}{2})$.

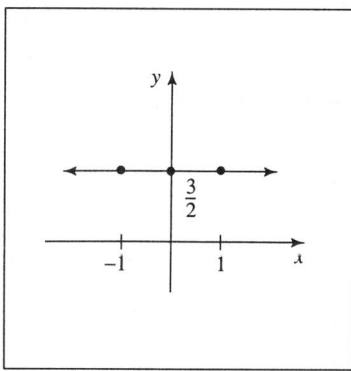

Figure 45

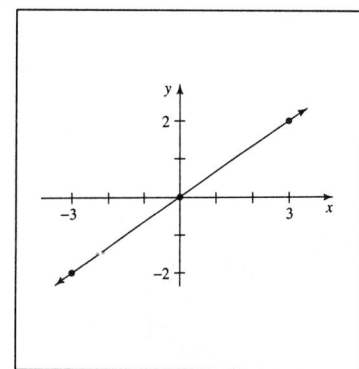

Figure 49

[49] $2x = 3y$ is equivalent to $y = \frac{2}{3}x$. If $x = 0$ then $y = \frac{2}{3} \cdot 0 = 0$ implying that $(0, 0)$ lies on the line. If $x = 3$ then $y = \frac{2}{3} \cdot 3 = 2$ implying that $(3, 2)$ lies on the line. If $x = -3$ then $y = \frac{2}{3} \cdot (-3) = -2$ implying that $(-3, -2)$ lies on the line.

Exercises 6.2

53 $x - y - 4 = y - x - 5$ simplifies to $2x = 2y - 1$ or if we solve for y, $y = x + \frac{1}{2}$.
If $x = 0$ then $y = \frac{1}{2}$; if $x = 1$ then $y = 1 + \frac{1}{2} = \frac{3}{2}$; and if $x = 2$ then $y = 2 + \frac{1}{2} = \frac{5}{2}$.
Therefore $(0, \frac{1}{2})$, $(1, \frac{3}{2})$, and $(2, \frac{5}{2})$ are three points on the line.

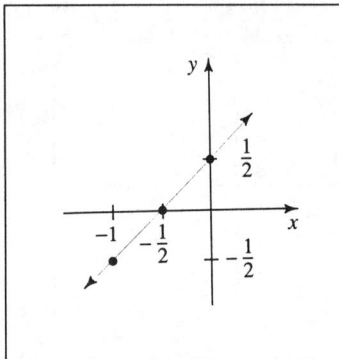

Figure 53

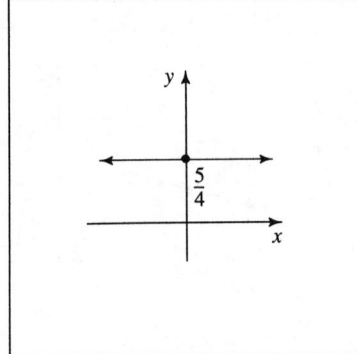

Figure 57

57 The equation $5 + 2y + 8x = 6y + 8x \Rightarrow 5 = 4y \Rightarrow y = \frac{5}{4}$. This is a horizontal line going through $(0, \frac{5}{4})$, $(1, \frac{5}{4})$, and $(2, \frac{5}{4})$.

61 To simplify $\frac{2x}{3} + \frac{y}{4} - \frac{1}{6} = \frac{y}{12} + \frac{3x}{2}$, multiply both sides by 12. Then
$12 \cdot \left(\frac{2x}{3} + \frac{y}{4} - \frac{1}{6} \right) = 12 \cdot \left(\frac{y}{12} + \frac{3x}{2} \right)$ which after multiplying by 12 becomes
$8x + 3y - 2 = y + 18x \Rightarrow 2y - 2 = 10x$. We can simplify further by multiplying both sides by $\frac{1}{2}$. This gives $y - 1 = 5x \Rightarrow y = 5x + 1$. Three points on this line are $(0, 1)$, $(-1, -4)$, and $(1, 6)$.

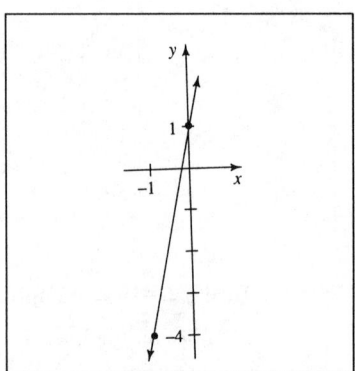

Figure 61

Exercises 6.2

Review Problems

69 $\frac{2-5}{1-3} = \frac{-3}{-2} = \frac{3}{2}$.

73 $\frac{1-6}{-\frac{1}{3}-\frac{1}{2}} = \frac{-5 \cdot 6}{(-\frac{1}{3}-\frac{1}{2}) \cdot 6} = \frac{-30}{6 \cdot (-\frac{1}{3}) - 6 \cdot \frac{1}{2}} = \frac{-30}{-2-3} = \frac{-30}{-5} = 6$.

Exercises 6.3

1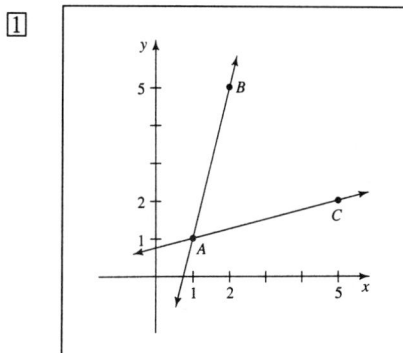

5 Point A is $(0, 0)$ and point B is $(3, 1)$.
The slope is $\frac{1-0}{3-0} = \frac{1}{3}$.

9 Point P is $(-2, 2)$ and point Q is $(2, -2)$.
The slope is $\frac{-2-2}{2-(-2)} = \frac{-4}{4} = -1$.

13 Given the points $(1, 3)$ and $(6, 5)$ the slope is not $\frac{5-3}{1-6} = -\frac{2}{5}$. The slope is $\frac{5-3}{6-1} = \frac{2}{5}$.
Note that the slope is also equal to $\frac{3-5}{1-6} = \frac{-2}{-5} = \frac{2}{5}$.

17 Given $(-4, 0)$ and $(0, 3)$ it is true that the slope is $\frac{3-0}{0-(-4)} = \frac{3}{4}$.

21 Given $(2, 5)$ and $(6, 13)$ the slope is $\frac{13-5}{6-2} = \frac{8}{4} = 2$.

25 The slope of the line through $(0, 0)$ and $(5, 5)$ is $\frac{5-0}{5-0} = \frac{5}{5} = 1$.

29 The slope of the line through $(-3, -6)$ and $(3, -2)$ is $\frac{-2-(-6)}{3-(-3)} = \frac{-2+6}{3+3} = \frac{4}{6} = \frac{2}{3}$.

Exercises 6.3

[33] Given $(1, \frac{11}{4})$ and $(4, 5)$ the slope is $\dfrac{5 - \frac{11}{4}}{4 - 1} = \dfrac{4 \cdot \left(5 - \frac{11}{4}\right)}{4 \cdot 3} = \dfrac{20 - 11}{12} = \dfrac{9}{12} = \dfrac{3}{4}$.

[37] The line determined by $(.5, 3)$ and $(.4, 2)$ has slope $\dfrac{2 - 3}{.4 - .5} = \dfrac{-1}{-.1} = \dfrac{1 \cdot 10}{.1 \cdot 10} = \dfrac{10}{1} = 10$.

[41] The line through $(1.4, .2)$ and $(6.2, 1.3)$ has slope $\dfrac{1.3 - .2}{6.2 - 1.4} = \dfrac{1.1}{4.8} = .23$.

[45] To find a 2nd point on a line through the point $(3, 1)$ with slope 1, note that

slope $= \dfrac{\text{Change in } y}{\text{Change in } x} = 1 = \dfrac{1}{1}$. Therefore change y by 1 and change x by 1 and we get the point

$(3+1, 1+1) = (4, 2)$. So $(4, 2)$ also lies on this line.

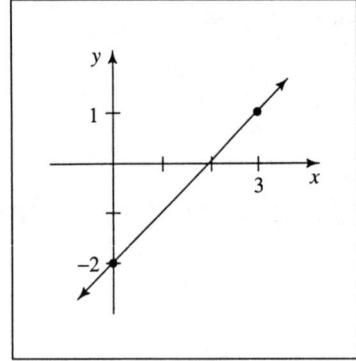

Figure 45

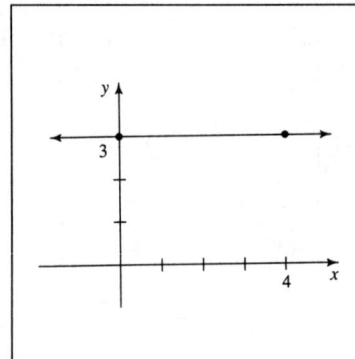

Figure 49

[49] The line through $(4, 3)$ with slope 0 is a horizontal line. The y value of any point on this line is 3. Thus a 2nd point is $(5, 3)$.

[53] To find 2 other points on the line through $(4, 0)$ with slope 2 note that

$2 = \dfrac{2}{1} = \dfrac{-2}{-1} = \dfrac{\text{Change in } y}{\text{Change in } x}$. If we add 2 to y and add 1 to x we get the point $(4+1, 0+2) =$

$(5, 2)$. If we add -2 to y and add -1 to x we get the point $(3, -2)$. Therefore

$(5, 2)$ and $(3, -2)$ are 2 other points on the line.

Exercises 6.3

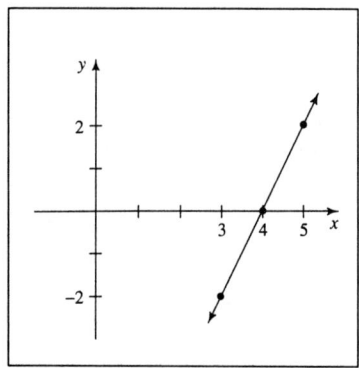

Figure 53

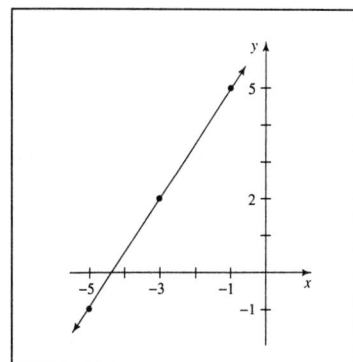

Figure 57

[57] To find 2 other points on the line through $(-3, 2)$ with slope $\frac{3}{2}$ note that $\frac{3}{2} = \frac{-3}{-2}$. So a 2nd point on the line is $(-3+2, 2+3) = (-1, 5)$. A 3rd point is $(-3+[-2], 2+[-3]) = (-5, -1)$.

[61] If $y = 3x - 4$ then 2 points on this line are $(0, -4)$ and $(1, -1)$. Therefore the slope is
$\frac{-1-(-4)}{1-0} = \frac{-1+4}{1} = 3$.

[65] Two points on the line $3x - 2y = 6$ are $(0, -3)$ and $(2, 0)$. Therefore the slope is
$\frac{0-(-3)}{2-0} = \frac{0+3}{2} = \frac{3}{2}$.

[69] $5y - 6x = 0$ is equivalent to $y = \frac{6}{5}x$. Therefore 2 points on this line are $(0, 0)$ and $(5, 6)$ and the slope is $\frac{6-0}{5-0} = \frac{6}{5}$.

[73] The slope of the line containing the points $(1, 5)$ and $(3, 9)$ is $\frac{9-5}{3-1} = \frac{4}{2} = 2$.
If $(x, -5)$ lies on this line then the slope from $(1, 5)$ to $(x, -5)$ must be 2. Therefore
$\frac{-5-5}{x-1} = 2 \Rightarrow \frac{-10}{x-1} = 2$. If we multiply both sides by $x - 1$ we get $-10 = 2(x-1) \Rightarrow -10 = 2x - 2 \Rightarrow -8 = 2x \Rightarrow x = -4$.

[77] If a line containing the points $(x, 10)$ and $(3x, 16)$ has slope $\frac{3}{4}$ then $\frac{16-10}{3x-x} = \frac{3}{4} \Rightarrow \frac{6}{2x} = \frac{3}{4}$. Cross-multiplying gives $6x = 24$ which means $x = 4$.

Page 115

Exercises 6.3

Review Problems

[85] If $y - 5 = -3(x - 3)$ then $y - 5 = -3x + 9$ or $y = -3x + 14$.

Exercises 6.4

[1] Use the point-slope formula $y - a = m(x - b)$ where m is the slope and (a, b) is a point on the line. Since the slope is 4 and the line goes through (2, 1), its equation is $y - 2 = 4(x - 2)$.

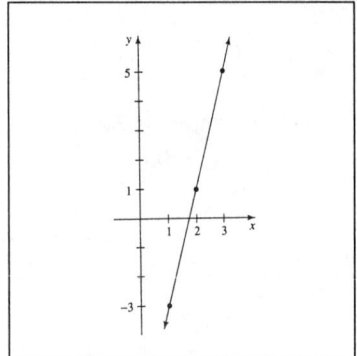

Figure 1

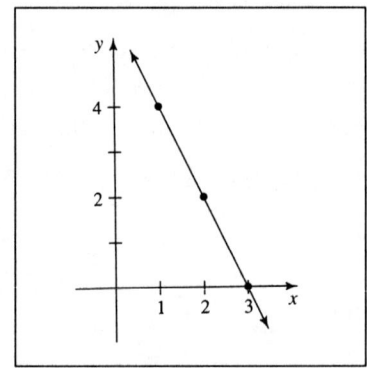

Figure 5

[5] If the slope is -2 and the line contains the point (0, 6), its equation is given by $y - 6 = -2(x - 0)$ or more simply $y - 6 = -2x$.

[9] If the line goes through (1, 1) and (3, 7) then $m = \dfrac{7-1}{3-1} = \dfrac{6}{2} = 3$. Then select one of the 2 points, say (1, 1), and use the point-slope formula $y - b = m(x - a)$. Therefore the equation is $y - 1 = 3(x - 1)$.

[13] The line through (6, 2) and (6, -4) is a vertical line since both points have the same x-coordinate 6. Therefore its equation is given by $x = 6$.

17 Use the slope-intercept form of the line $y = mx + b$ where m is the slope and b is the y-intercept. Since $(0, 4)$ is on the line we know the y-intercept is 4. Therefore $b = 4$. Since $m = 2$ the equation is $y = 2x + 4$.

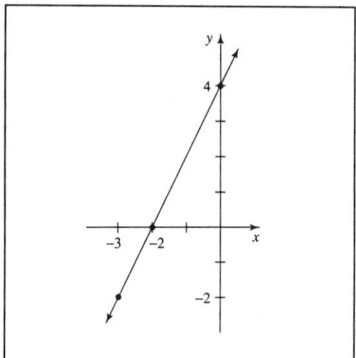

Figure 17

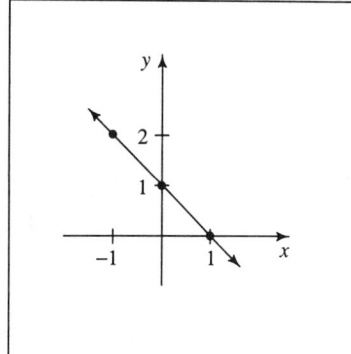

Figure 21

21 If $(-3, 4)$ is on the line and $m = -1$ then the equation is given by $y - 4 = -1(x - [-3])$ or $y - 4 = -(x + 3)$. Then $y = -x - 3 + 4$ or $y = -x + 1$ which is the slope-intercept form of the line.

25 If the line contains $(1, 3)$ and $(2, 5)$, $m = \dfrac{5-3}{2-1} = 2$. Then the equation is given by $y - 3 = 2(x - 1)$ or $y - 3 = 2x - 2$. Add 3 to both sides to obtain the slope-intercept form which is $y = 2x + 1$.

29 The line through $(4, 6)$ and $(-4, 1)$ has slope $m = \dfrac{1-6}{-4-4} = \dfrac{-5}{-8} = \dfrac{5}{8}$. Its equation is $y - 1 = \dfrac{5}{8}(x - [-4]) \Rightarrow y - 1 = \dfrac{5}{8}(x + 4) \Rightarrow y - 1 = \dfrac{5}{8}x + \dfrac{5}{8} \cdot 4$. Add $1 = \dfrac{8}{8}$ to both both sides to put the equation in slope intercept form. This gives $y = \dfrac{5}{8}x + \dfrac{20}{8} + \dfrac{8}{8} \Rightarrow y = \dfrac{5}{8}x + \dfrac{28}{8} \Rightarrow y = \dfrac{5}{8}x + \dfrac{7}{2}$.

33 The line through $(0, -2)$ and $(1, 0)$ has slope $m = \dfrac{0-(-2)}{1-0} = 2$. Since $(0, -2)$ is on the line we know that $b = -2$. Therefore, using $y = mx + b$, the equation is $y = 2x - 2$.

[37] The line through $(-2, 3)$ and $(-2, -5)$ is a vertical line since both points have the same x-coordinate -2. Hence the equation is $x = -2$.

[41] The slope of the line through $(0, 0)$ and $(2, \frac{6}{5})$ is $m = \dfrac{\frac{6}{5} - 0}{2 - 0} = \frac{6}{5} \div \frac{2}{1} = \frac{6}{5} \cdot \frac{1}{2} = \frac{6}{10} = \frac{3}{5}$.

We know b is 0 since the line intersects the y-axis at $(0, 0)$. Therefore the equation is $y = \frac{3}{5}x$.

[45] If $(6.2, 2.1)$ and $(7.8, 4.6)$ are two points on the line then $m = \dfrac{4.6 - 2.1}{7.8 - 6.2} = \dfrac{2.5}{1.6} \approx 1.56$.

Therefore the equation is $y - 2.1 = 1.56(x - 6.2)$.

[49] Put the equation $y - 3 = 2(x - 1)$ into slope-intercept form by solving for y.

$y - 3 = 2(x - 1) \Rightarrow y - 3 = 2x - 2 \Rightarrow y = 2x + 1$.

Thus we see that the slope is 2 and the y-intercept is 1. To find the x-intercept make $y = 0$.

Then $0 = 2x + 1 \Rightarrow 2x = -1 \Rightarrow x = -\frac{1}{2}$. Therefore

a) The slope is 2   b) The y-intercept is 1   c) The x-intercept is $-\frac{1}{2}$

[53] $2x - 5y = 10$ is equivalent to $5y = 2x - 10$. After we multiply both sides by $\frac{1}{5}$ we get

$y = \frac{2}{5}x - 2$. Don't forget to multiply -10 by $\frac{1}{5}$. This means $m = \frac{2}{5}$ and $b = -2$.

Let $y = 0$. Then $0 = \frac{2}{5}x - 2 \Rightarrow 2 = \frac{2}{5}x \Rightarrow x = \frac{5}{2} \cdot 2 = 5$. Therefore

a) The slope is $\frac{2}{5}$ b) The y-intercept is -2 c) The x-intercept is 5

[57] The lines given by $y = 2x + 4$ and $y = 2x - 3$ have the same slope of 2. If 2 lines have the same slope then they are parallel lines.

[61] The lines $y + 1 = 6(x - 1)$ and $y - 1 = 6(x + 3)$ each have slope 6. Therefore these lines are parallel.

[65] $x = -3$ and $x = 5$ are both vertical lines. They are parallel because two distinct vertical lines are parallel.

Exercises 6.4

[69] a) The slope of the line $y = 2x + 1$ is 2. Any line parallel must have slope 2. Therefore the parallel line through the point (2, 4) has equation $y - 4 = 2(x - 2)$.

b) The line perpendicular to $y = 2x + 1$ must have a slope which is the negative reciprocal of 2. The negative reciprocal of 2 is $-\frac{1}{2}$. Therefore the perpendicular line through (2, 4) has the equation $y - 4 = -\frac{1}{2}(x - 2)$.

[73] Note that the line $x = 1$ is vertical. So a line parallel must be vertical and a line perpendicular must be horizontal.

a) The line parallel through (3, 5) must be a vertical line where each point has x-coordinate 3. Therefore its equation is given by $x = 3$.

b) The line perpendicular through (3, 5) must be a horizontal line where each point has y-coordinate 5. The equation of this line is $y = 5$.

[77] To find the slope of $\frac{3}{4}x - \frac{1}{2}y = 1$ solve for y to put the equation in the slope-intercept form. The equation is equivalent to $\frac{3}{4}x - 1 = \frac{1}{2}y$. When we multiply both sides by 2 we get $y = 2 \cdot \frac{3}{4}x - 2 \cdot 1 \Rightarrow y = \frac{3}{2}x - 2$. This means $m = \frac{3}{2}$.

a) The line parallel must have slope $\frac{3}{2}$. Thus the parallel line through $(\frac{2}{5}, -3)$ has equation $y - (-3) = \frac{3}{2}\left(x - \frac{2}{5}\right)$ or better $y + 3 = \frac{3}{2}\left(x - \frac{2}{5}\right)$.

b) The perpendicular line has slope equal to the negative reciprocal of $\frac{3}{2}$ which is $-\frac{2}{3}$. Its equation is given by $y + 3 = -\frac{2}{3}\left(x - \frac{2}{5}\right)$.

[81] The slope of the line through A(2, 2) and B(−4, −1) is $m = \frac{-1-2}{-4-2} = \frac{-3}{-6} = \frac{1}{2}$. The parallel line has slope $\frac{1}{2}$ and the perpendicular line has slope $-\frac{2}{1} = -2$. Therefore

a) The parallel line through P(6, 0) is $y - 0 = \frac{1}{2}(x - 6)$ or more simply $y = \frac{1}{2}x - 3$.

b) The perpendicular line through P(6, 0) is $y - 0 = -2(x - 6)$ or $y = -2x + 12$.

[85] The slope of the line from A(3, −3) to B(−4, 4) is $\frac{4-(-3)}{-4-3} = \frac{4+3}{-7} = \frac{7}{-7} = -1$. The parallel line has slope -1 and the perpendicular line has slope 1.

a) The line parallel through P(5, 5) is $y - 5 = -1(x - 5) \Rightarrow y - 5 = -x + 5 \Rightarrow y = -x + 10$.

b) The line perpendicular through P(5, 5) is $y - 5 = 1(x - 5)$ or more simply $y = x$.

89. A firm's profit/loss equation is $y = 4x - 260$.

   a) The break-even point occurs when the profit is 0. Since y is the profit, make $y = 0$. Then $0 = 4x - 260 \Rightarrow 260 = 4x \Rightarrow x = \frac{260}{4} = 65$. So the break-even point is 65.

   b) To find the profit when 140 items are sold, make $x = 140$. Then $y = 4x - 260 = 4 \times 140 - 260 = 560 - 260 = 300$. Therefore the profit is $300.

   c) If 0 items are sold make $x = 0$. Then $y = 4 \times 0 - 260 = 0 - 260 = -260$. So there is a loss of $260.

93. The triangle with vertices A(0, 0), B(2, 1), and C(0, 5) is a right triangle if 2 of the sides are perpendicular. Two lines are perpendicular if their slopes are negative reciprocals. The slope from A to B is $\frac{1-0}{2-0} = \frac{1}{2}$ and the slope from B to C is $\frac{5-1}{0-2} = \frac{4}{-2} = -2$. Since $\frac{1}{2}$ and -2 are negative reciprocals, these sides are perpendicular and the triangle must be a right triangle.

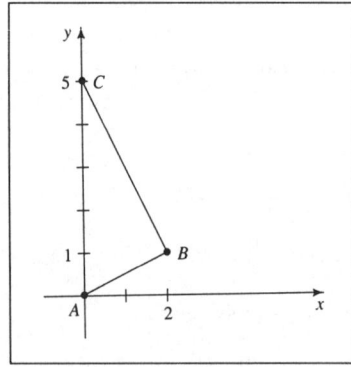

Figure 93

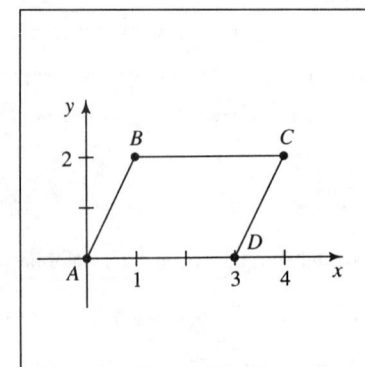

Figure 97

97. The quadrilateral with vertices A(0, 0), B(1, 2), C(4, 2), and D(3, 0) is a parallelogram if the opposite sides are parallel. Since sides AB and CD are opposite, we must check to see if the lines are parallel. The line through A and B has slope $\frac{2-0}{1-0} = 2$ and the line through

Exercises 6.4

C and D is $\frac{0-2}{3-4} = \frac{-2}{-1} = 2$. Since the slopes are equal the lines are parallel.

The sides BC and AD are opposite sides and we must check if these lines are parallel. The side from B to C has slope $\frac{2-2}{4-1} = 0$ and the side from A to D has slope $\frac{0-0}{3-0} = 0$. Therefore these sides are parallel and the figure is a parallelogram.

Review Problems

$\boxed{105}$ $x > 2$

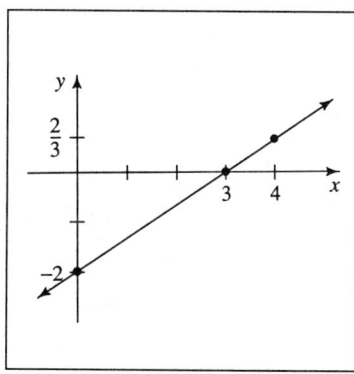

$\boxed{109}$ The line $2x - 3y = 6$ has x-intercept 3 and y-intercept -2. Therefore the line goes through $(3, 0)$ and $(0, -2)$. The line also goes through $(-3, -4)$.

Exercises 6.5

[1] To determine the correct half-plane for $y \leq x+2$ use $(0, 0)$ as a test point. $(0, 0)$ satisfies $y \leq x+2$ because $0 \leq 0+2$ is a true statement. Therefore we want the half-plane which includes the point $(0, 0)$, that is the half-plane which goes downward.

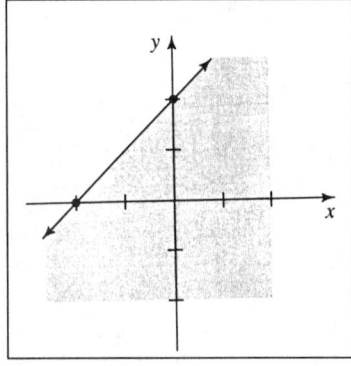

Figure 1

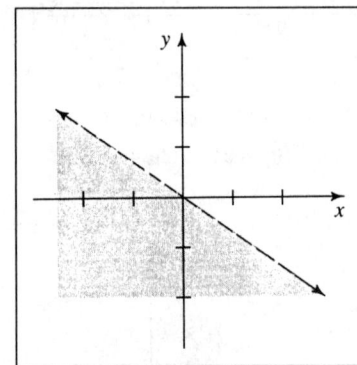

Figure 5

[5] To determine the half-plane of $x+2y < 0$ use a point such as $(1, 0)$. $(1, 0)$ does not satisfy $x+2y < 0$ because $1+2 \times 0 < 0$ or $1 < 0$ is false. Therefore the correct half-plane is the one which does not include the point $(1, 0)$, that is the one opening downward.

[9] To graph $y \leq x+5$ first graph the boundary line $y = x+5$ as a solid line since the graph includes this line. This line goes through the points $(0, 5)$, $(5, 0)$, and $(-2, 3)$. To determine which half-plane, use $(0, 0)$ as a test point. $(0, 0)$ satisfies $y \leq x+5$ since $0 \leq 0+5$ is true. Therefore we select the half-plane opening downward which includes the point $(0, 0)$.

Exercises 6.5

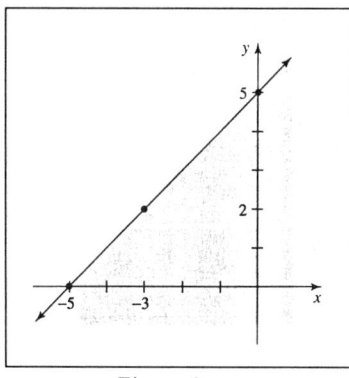

Figure 9

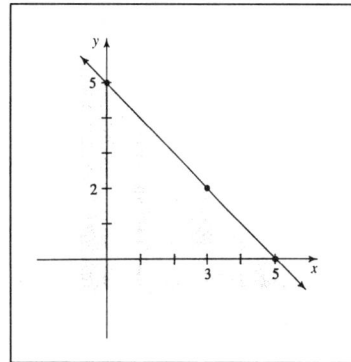

Figure 13

13. To graph $x + y \leq 5$, graph the line $x + y = 5$ as a solid line. This line passes through points $(5, 0)$, $(0, 5)$, and $(2, 3)$. Then determine which half-plane by using $(0, 0)$ as a test point in $x + y \leq 5$. $(0, 0)$ satisfies $x + y \leq 5$ implying that the half-plane must include $(0, 0)$. Thus we select the half-plane going downward.

17. The half-plane given by $y \geq x$ has boundary line $y = x$. Graph $y = x$ as a solid line since the graph includes the boundary line. Then use $(0, 1)$ as a test point in $y \geq x$. Since $1 \geq 0$ is true, the half-plane opens upward because this includes the point $(0, 1)$.

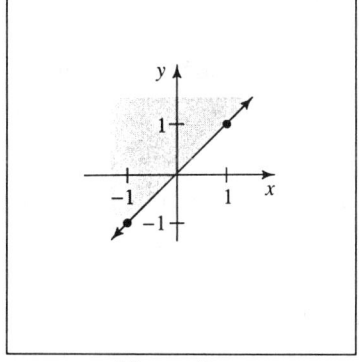

Figure 17

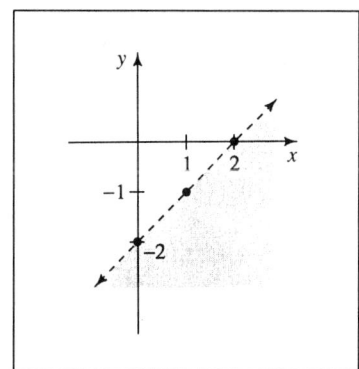

Figure 21

Page 123

21 The half-plane $y - x < -2$ is bounded by, but does not include the line $y - x = -2$. Therefore graph the line $y - x = -2$ as a dotted line. This line passes through $(0, -2)$ and $(2, 0)$. Use $(0, 0)$ as a test point in $y - x < -2$. $(0, 0)$ does not satisfy $y - x < -2$ since $0 - 0 < -2$ or $0 < -2$ is false. Thus the half-plane opens downward since $(0, 0)$ is not on the graph.

25 To graph $4x - 3y > 12$, 1st graph the line $4x - 3y = 12$ as a dotted line because the boundary line is not included in the graph. Then use $(0, 0)$ as a test point in $4x - 3y > 12$. It does not satisfy the inequality because $0 - 0 > 12$ is false. Hence the half-plane opens downward.

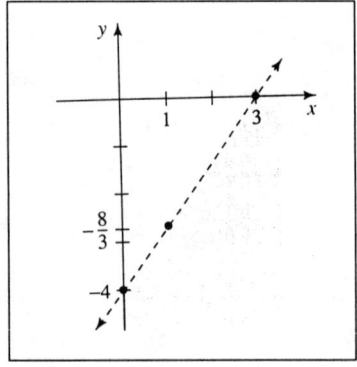

Figure 25

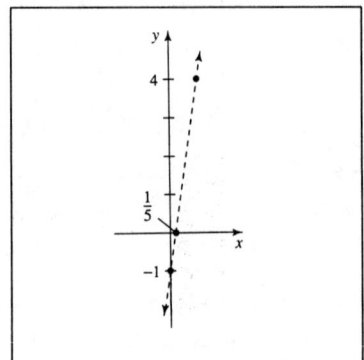

Figure 29

29 The half-plane given by $y < 5x - 1$ is bounded by, but does not include the line $y = 5x - 1$. Graph this which passes through $(0, -1)$ and $(1, 4)$ as a dotted line. The half-plane opens downward because $(0, 0)$ does not satisfy $y < 5x - 1$ and $(0, 0)$ can not be part of the graph.

33 The half-plane $3x < 4$ is bounded by, but does not include the vertical line $3x = 4$ which is more simply written as $x = \frac{4}{3}$. If we use $(0, 0)$ as a test point we see that $0 < 4$ is true implying that the half-plane must open to the left.

Exercises 6.5

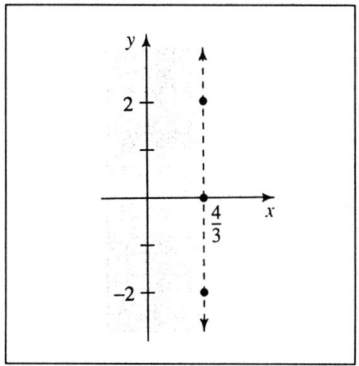

Figure 33

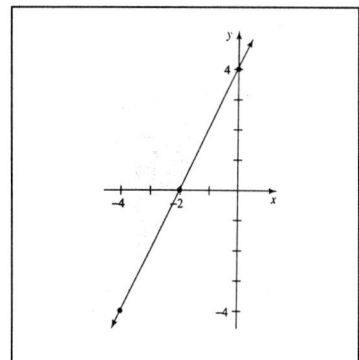

Figure 37

37 $\frac{x}{2} - \frac{y}{4} \leq -1$ is bounded by, and includes the line $\frac{x}{2} - \frac{y}{4} = -1$. $\frac{x}{2} - \frac{y}{4} = -1 \Rightarrow$
$4 \cdot \left(\frac{x}{2} - \frac{y}{4}\right) = 4 \cdot (-1) \Rightarrow 2x - y = -4$. This line passes through $(0, 4)$ and
$(-2, 0)$. Checking $(0, 0)$ we see this does not satisfy $\frac{x}{2} - \frac{y}{4} \leq -1$ since $0 - 0 \leq -1$ is false.
Therefore the half-plane opens upward because $(0, 0)$ is not part of the graph.

41 a) The graph of $x \leq 4$ on the number line is

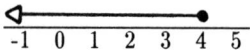

b) The graph of $x \leq 4$ in the xy-plane is a half-plane opening to the left bounded by the vertical line $x = 4$.

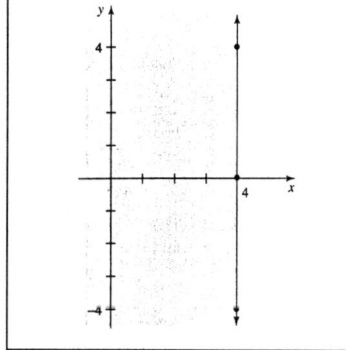

Page 125

Exercises 6.5

45 a) The graph of $-6 \leq x < 1$ on the number is

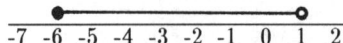

b) The graph of $-6 \leq x < 1$ in the xy-plane is the set of all points where the x-coordinate is between -6 and 1, including -6 but not 1. This is the region bounded by the vertical lines $x = -6$ and $x = 1$. So 1st draw $x = -6$ as a solid line and $x = 1$ as a dotted line.

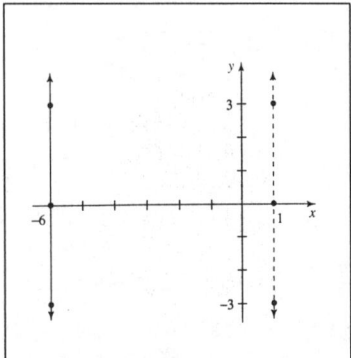

Figure 45

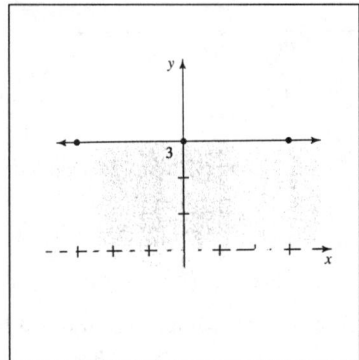

Figure 49

49 The graph of $0 < y \leq 3$ in the xy-plane is the set of points where the y-coordinate is between 0 and 3, including 3 but not 0. This region is bounded by the horizontal lines $y = 0$ and $y = 3$. Draw $y = 3$ as a solid line because the graph includes this line. The line $y = 0$ is the x-axis which is not part of the graph.

53 The set of points (x, y) such that $x \geq -4$ and $y > 1$ is the intersection of the two half-planes $x \geq -4$ and $y > 1$. $x \geq -4$ is the half-plane bounded by the vertical line $x = -4$ opening to the right, including the line. The half-plane $y > 1$ is the half-plane bounded by the horizontal line $y = 1$ opening upward, but not including this line. So draw $x = -4$ as a solid line and $y = 1$ as a dotted line.

Exercises 6.5

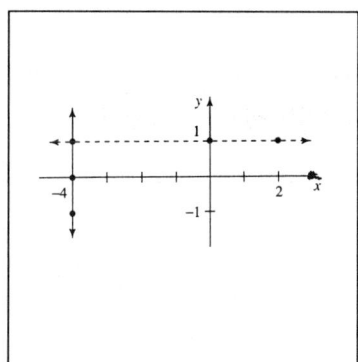

Figure 53

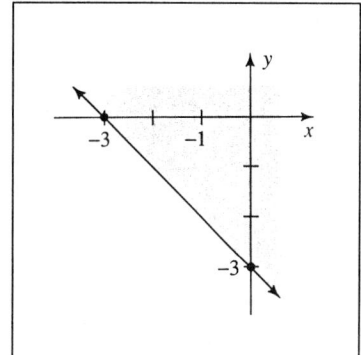

Figure 57

[57] $2(y-3) < 2(x+2y)$ is equivalent to $2y - 6 < 2x + 4y$ or $-6 < 2x + 2y$. This can be written as $2x + 2y > -6$ and if we divide both sides by 2, $x + y > -3$. The graph is the half-plane bounded by, but not including the line $x + y = -3$. This line passes through $(0, -3)$ and $(-3, 0)$. The half-plane opens to the right because $(0, 0)$ satisfies $x + y > -3$.

Chapter Review Exercises

1.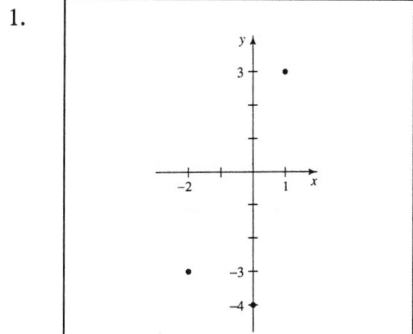

5. If $x = 0$ then $y = 2 \cdot 0 - 4 = -4$. Hence $(0, -4)$ is a solution.
If $x = 3$ then $y = 2 \cdot 3 - 4 = 2$. Hence $(3, 2)$ is a solution.
If $y = -14$ then $-14 = 2x - 4 \Rightarrow -10 = 2x \Rightarrow x = -5$.
Therefore $(-5, -14)$ is a solution.

Chapter 6 Review Exercises and Test

9. Given $y = 4x - 3$, if $y = 0$ then $0 = 4x - 3$ or $4x = 3$ which means $x = \frac{3}{4}$ is the x-intercept. If $x = 0$ then $y = 4 \cdot 0 - 3 = -3$ is the y-intercept.

13. Given $y = 4x - 3$, if $x = 0$ then $y = 4 \cdot 0 - 3 = -3$. Hence $(0, -3)$ is one point on the line. If $x = 1$ then $y = 4 \cdot 1 - 3 = 1$. So $(1, 1)$ is a point on the graph. Similarly a 3rd point on the graph is $(2, 5)$.

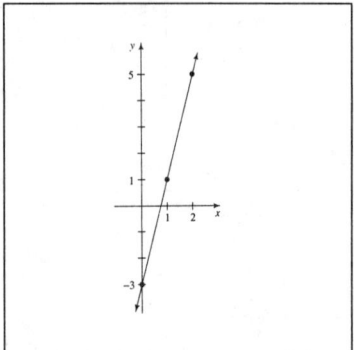

Figure 13

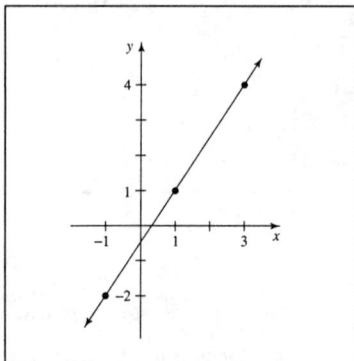

Figure 17

17. The line $2y + 1 = 3x$ can be rewritten as $y = \frac{3x - 1}{2}$. Then $x = 1$ implies that $y = \frac{3-1}{2} = 1$. Therefore $(1, 1)$ is a point on the line. Similarly $(3, 4)$ and $(-1, -2)$ are also on the line.

21. The slope of the line through $(2, 3)$ and $(-1, -3)$ is $\frac{3 - (-3)}{2 - (-1)} = \frac{6}{3} = 2$.

25. The slope of the line determined by $\left(3, \frac{1}{2}\right)$ and $\left(\frac{1}{4}, -2\right)$ is $\frac{-2 - \frac{1}{2}}{\frac{1}{4} - 3} = \frac{\left(-2 - \frac{1}{2}\right) \cdot 4}{\left(\frac{1}{4} - 3\right) \cdot 4} = \frac{-8 - 2}{1 - 12} = \frac{-10}{-11} = \frac{10}{11}$.

29. If a firm's profit/loss equation is $y = 6x - 900$ then its break-even point occurs when $y = 0$. Therefore $0 = 6x - 900 \Rightarrow 6x = 900 \Rightarrow x = \frac{900}{6} = 150$. So the break-even point is 150.

33. The slope of the line through $(2, 3)$ and $(4, 1)$ is $\frac{1-3}{4-2} = -1$. Then, using the point $(4, 1)$, the equation is $y - 1 = -1(x - 4) \Rightarrow y - 1 = -x + 4 \Rightarrow y = -x + 5$.

Chapter 6 Review Exercises and Test

37. The slope from $(1, -2)$ to $(-1, 6)$ is $\frac{6-(-2)}{-1-1} = \frac{8}{-2} = -4$. Using $(-1, 6)$, the equation is
$y - 6 = -4(x - [-1]) \Rightarrow y - 6 = -4(x + 1)$.

41. The line $2x - 3y = 3$ is equivalent to $y = \frac{2}{3}x - 1$. Therefore its slope is $\frac{2}{3}$.
The line $3x + 2y = 8$ is equivalent to $y = -\frac{3}{2}x + 4$. Therefore its slope is $-\frac{3}{2}$.
Since $\frac{2}{3}$ and $\frac{-3}{2}$ are negative reciprocals, the lines are perpendicular.

45. The graph of $y < -x + 2$ is a half-plane bounded by, but not including the line $y = -x + 2$. So first graph $y = -x + 2$ as a dotted line. Then we can see that the half-plane must open downward because $(0, 0)$ satisfies the inequality $y < -x + 2$.

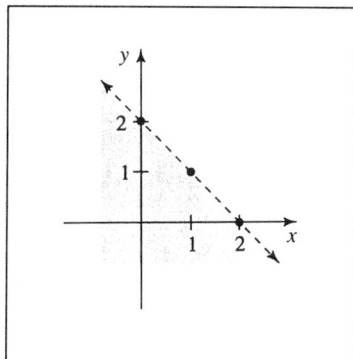

Figure 45

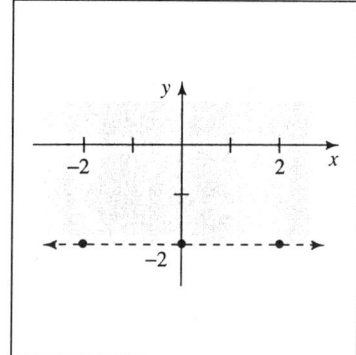

Figure 49

49. The graph of $y > -2$ is the half-plane bounded by, but not including the horizontal line $y = -2$. It opens upward. So graph the line $y = -2$ as a dotted line and shade the region above.

Chapter Test

1. If $y = 3x - 2$ and $x = 0$ then $y = 0 - 2 = -2$ implying that $(0, -2)$ is a solution.
   If $x = -4$ then $y = -12 - 2 = -14$. Therefore $(-4, -14)$ is a solution.
   If $y = 13$ then $13 = 3x - 2 \Rightarrow 15 = 3x$. Then $x = 5$ implying that $(5, 13)$ is a solution.

5. Given $y = 3x - 2$, if $x = 0$ then $y = 3 \cdot 0 - 2 = -2$. Therefore $(0, -2)$ is a point on the line. Similarly $(1, 1)$ and $(2, 4)$ lie on the line.

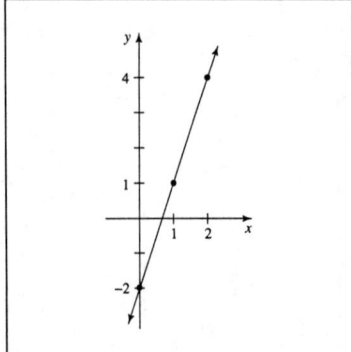

Figure 5

9. The slope is $\dfrac{-10 - 2}{-2 - 1} = \dfrac{-12}{-3} = 4$.

13. The equation is given by $y - 4 = 3(x - [-1])$ or $y - 4 = 3(x + 1)$.

17. If a line contains the points $(-4, -1)$ and $(-4, 6)$ it must be the vertical line where each point has x-coordinate -4. Therefore its equation is given by $x = -4$.

21. Since $2x - y = 3$ can be rewritten as $y = 2x - 3$ we see its slope is 2. Therefore any line parallel must have slope 2 and any line perpendicular has slope $-\dfrac{1}{2}$.
    a) The line parallel through the point $(4, -1)$ has the equation $y - (-1) = 2(x - 4)$ or $y + 1 = 2(x - 4)$.
    b) The line perpendicular through the point $(4, -1)$ has the equation $y + 1 = -\dfrac{1}{2}(x - 4)$.

Chapter 6 Review Exercises and Test

25. The graph of $2x + 5y > 0$ is the half-plane above the line $2x + 5y = 0$.

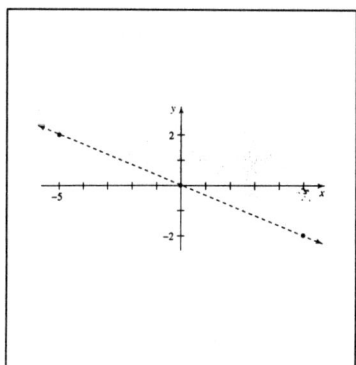

Cumulative Review

1. $6 + 4|-5| = 6 + 4 \times 5 = 6 + 20 = 26$.

5. $-5^2 = -(5^2) = -25$.

9. $a^2(2a^6)^3 = a^2 \cdot 2^3(a^6)^3 = 8a^2 a^{18} = 8a^{20}$.

13. $8 + 2(3x - 5) = 8 + 6x - 10 = 6x - 2$.

17. $\dfrac{x^2 - 6x + 2}{2x} = \dfrac{x^2}{2x} - \dfrac{6x}{2x} + \dfrac{2}{2x} = \dfrac{x}{2} - 3 + \dfrac{1}{x}$.

21. $\dfrac{2}{x^2 - x} - \dfrac{x-3}{x^2 - 1} = \dfrac{2}{x(x-1)} - \dfrac{x-3}{(x+1)(x-1)} = \dfrac{2(x+1) - x(x-3)}{x(x+1)(x-1)} = \dfrac{2x + 2 - x^2 + 3x}{x(x+1)(x-1)} =$

$\dfrac{-x^2 + 5x + 2}{x(x+1)(x-1)}$.

25. $5y - (y - 8) = 10y$
$5y - y + 8 = 10y$
$4y + 8 = 10y$
$8 = 6y$
$y = \dfrac{8}{6} = \dfrac{4}{3}$.

29. $m = \dfrac{y - b}{x}$
$mx = y - b$
$mx + b = y$
$y = mx + b$.

Cumulative Review: Chapters 1-6

33. $\frac{1}{S} = \frac{1}{E} - \frac{1}{P}$

$SEP\frac{1}{S} = SEP \cdot \left(\frac{1}{E} - \frac{1}{P}\right)$

$EP = SP - SE$

$SE = SP - EP$

$SE = P(S - E)$

$P = \frac{SE}{S - E}$.

37. $ax - 3y - 3x - ay$ is not factorable.

41. If $x = -2$ then $x^2 - 3x + 4 =$
$(-2)^2 - 3(-2) + 4 = 4 + 6 + 4 = 14$.

45. The domain of $\frac{x}{x^2 + 9}$ is all real numbers because $x^2 + 9$ is never 0.

49. The rational numbers of $\{-1, 0, 1, \sqrt{2}, 3.7, 3.\overline{7}, 3.7172737475...\}$ are $-1, 0, 1, 3.7,$ and $3.\overline{7}$.

53. The slope is $\frac{4-3}{0-2} = \frac{1}{-2} = -\frac{1}{2}$.

57. The graph of $x - 2y > 4$ is a half-plane bounded by, but not including the line $x - 2y = 4$. The half-plane opens downward because (0, 0) does not satisfy the inequality.

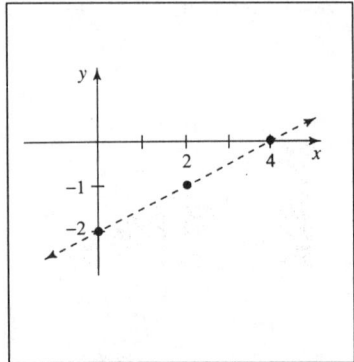

Figure 57

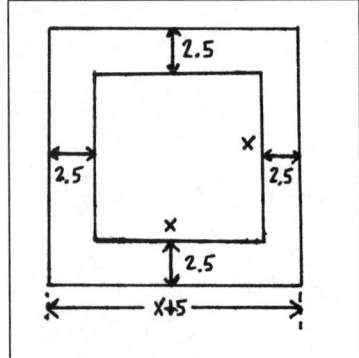

Figure 61

61. Let x be the length of the side of the original lawn. Then the length of the side of the new lawn is $x+5$ and the new lawn has area $(x+5)^2$. If this area is 4 times the original area x^2 then $(x+5)^2 = 4x^2 \Rightarrow x^2 + 10x + 25 = 4x^2 \Rightarrow 3x^2 - 10x - 25 = 0$.
Factoring the left side gives $(3x+5)(x-5) = 0 \Rightarrow x = 5$ or $-\frac{5}{3}$.
Therefore we see that the original side is 5 meters.

65. If x is the cost of the calculator then $x + .06x = 25.96$ which means $1.06x = 25.96$. Therefore $x = \frac{25.96}{1.06} = 24.49$. So she paid $24.49 for the calculator and $.06 \times 24.49 = \$1.47$ in tax.

69. Let x be Steve's age and $x+3$ Paula's age. Then $x + 6 + x + 3 + 6 = 55 \Rightarrow 2x + 15 = 55 \Rightarrow 2x = 40 \Rightarrow x = 20$. Therefore Steve is 20 and Paula is 23.

Chapter 7: Linear Systems

Exercises 7.1

[1] To find the point of intersection of the lines $y = 3x - 4$ and $y = -4x + 7$ using a TI-81 calculator, press ZOOM 6 to set the calculator to the standard graphing screen. Then press Y= and make $y_1 = 3x - 4$ and $y_2 = -4x + 7$. Press GRAPH to draw the two graphs. To find the point of intersection, press TRACE and use the arrow keys to move the trace bug to the point of intersection. Press ZOOM 2 ENTER and again press TRACE and put the trace bug on the point of intersection. You can repeat this process until the coordinates are accurate to the nearest hundredth.

[5] The ordered pair $(-4, -1)$ is not a solution to the system $\begin{array}{l} 3y - 2x = 5 \\ x = 2y - 3 \end{array}$ despite the fact that it satisfies the 1st equation: $3 \cdot (-1) - 2 \cdot (-4) = -3 + 8 = 5$. It does not satisfy the 2nd equation: $-4 = 2 \cdot (-1) - 3$ is false because $-4 \neq -5$.

[9] $(6, 0)$ is not a solution to $\begin{array}{l} x + 3y = 6 \\ 3y = -x + 3 \end{array}$ because it does not satisfy the 2nd equation: $3 \cdot (0) = -6 + 3$ is false since $0 \neq -3$.

[13] The line $x = -2$ is a vertical line and $y = 3$ is a horizontal line. They intersect at $(-2, 3)$. Therefore the solution is $(-2, 3)$.

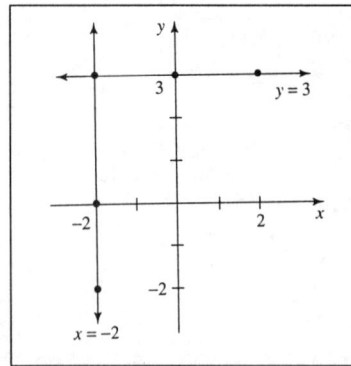

Figure 13

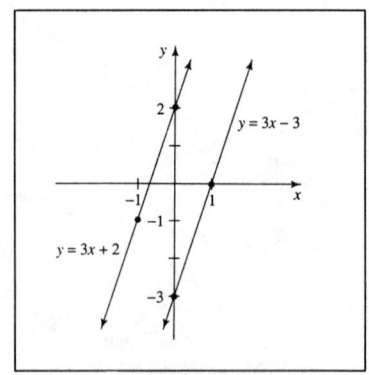

Figure 17

Exercises 7.1

[17] The lines $y = 3x + 2$ and $y = 3x - 3$ are parallel since both lines have slope 3. Therefore the lines do not intersect and the equations are inconsistent.

[21] $x = 2y + 4$ can be rewritten as $y = \frac{x-4}{2}$ and $-x + 2y = -4$ can be rewritten as $y = \frac{x-4}{2}$. Hence the lines are the same so the equations are dependent. Note that the points $(0, -2)$ and $(4, 0)$ lie on both lines.

[25] $8x - 4y = 12$ is equivalent to $y = \frac{8x - 12}{4}$ or $y = 2x - 3$ and $-6x + 3y = -9$ is equivalent to $y = \frac{6x - 9}{3}$ or $y = 2x - 3$. Therefore the lines are the same and the equations are dependent. $(0, -3)$ and $(1, -1)$ lie on both lines.

[29] The lines $y = 4x + 1$ and $y = 4x - 1$ are parallel because both have slope 4. Therefore the equations are inconsistent.

[33] The lines $x = 4$ and $y = -2x + 5$ intersect at $(4, -3)$.

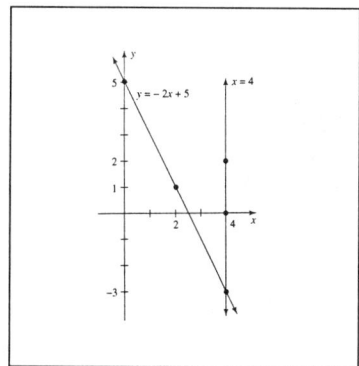

Figure 33

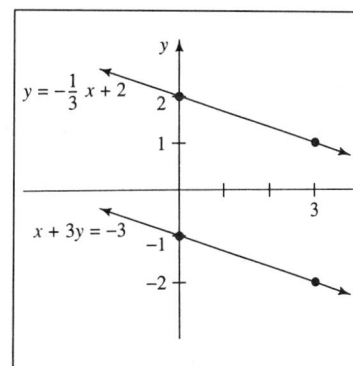

Figure 37

[37] $x + 3y = -3$ can be rewritten as $3y = -x - 3$ or $y = -\frac{1}{3}x - 1$. This means the slope is $-\frac{1}{3}$. $y = -\frac{1}{3}x + 2$ also has slope $-\frac{1}{3}$. Therefore the lines are parallel and the equations are inconsistent.

Exercises 7.1

41 To see if the lines $y = 1$, $x = 4$, and $y = 2x - 1$ determine a right triangle, we must determine if one of the angles is a right angle. The line $y = 1$ is a horizontal line and the line $x = 4$ is a vertical line. These lines intersect at a right angle. Therefore the triangle is a right triangle.

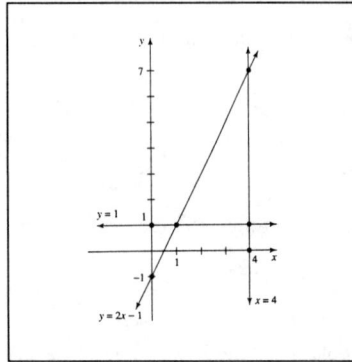

Figure 41

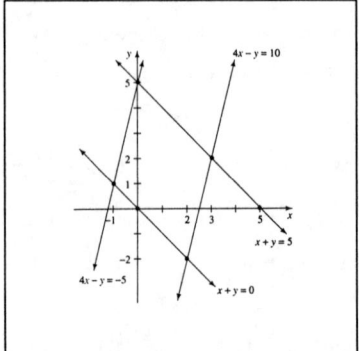

Figure 45

45 The vertices of the parallelogram formed by the lines $4x - y = -5$, $x + y = 5$, $x + y = 0$, and $4x - y = 10$ are $P_1 = (-1, 1)$, $P_2 = (0, 5)$, $P_3 = (2, -2)$, and $P_4 = (3, 2)$.

The slope of the line segment from P_1 to P_2 is $\dfrac{5-1}{0-(-1)} = \dfrac{4}{1} = 4$.

The slope of the line segment from P_3 to P_4 is $\dfrac{2-(-2)}{3-2} = \dfrac{2+2}{1} = 4$.

The slope of the line segment from P_1 to P_3 is $\dfrac{-2-1}{2-(-1)} = \dfrac{-3}{3} = -1$.

The slope of the line segment from P_2 to P_4 is $\dfrac{2-5}{3-0} = \dfrac{-3}{2} = -1$.

Since opposite sides have the same slope, the opposite sides are parallel and the figure is a parallelogram.

Review Problems

53 $3x - y = 2$
$3x = y + 2$
$3x - 2 = y \Rightarrow y = 3x - 2$.

57 $6x - 2(3x - 4) = -16$
$6x - 6x + 8 = -16$
$8 = -16$. This is a false equation and the solution set is $\emptyset$.

Exercises 7.2

1. To solve $y = 3x + 4$ substitute $y = 3x + 4$ into the 2nd equation. This gives
$y = 2x - 1$
$3x + 4 = 2x - 1 \Rightarrow x = -5$. Substitute $x = -5$ into the 1st equation.
Then $y = 3 \times (-5) + 4 = -15 + 4 = -11$ and the solution is $(-5, -11)$. We only have to check this solution in the 2nd equation: $-11 = 2 \times (-5) - 1 = -10 - 1 = -11$.

5. To solve $2x - y = 30$ solve the 1st equation for y and substitute for y in the 2nd equation.
$5x - 2y = 20$
So $2x - y = 30 \Rightarrow y = 2x - 30$ and substituting for y in the 2nd equation gives
$5x - 2(2x - 30) = 20 \Rightarrow 5x - 4x + 60 = 20 \Rightarrow x = -40$ and $y = 2 \cdot (-40) - 30 = -110$.
Therefore the solution is $(-40, -110)$.

9. Given the system $2x + y = 4$ we see from eq 1 that $y = -2x + 4$. Then substituting
$6x + 3y = -2$
into eq 2 gives $6x + 3(-2x + 4) = -2 \Rightarrow 6x - 6x + 12 = -2 \Rightarrow 12 = -2$.
This is a false equation so the equations are inconsistent. Therefore there are no solutions.

13. To solve $2x - 4y = 6$ solve the 2nd equation for x which gives $2y + 3 = x$ or $x = 2y + 3$.
$-x + 2y = -3$
Then substitute this into the 1st equation: $2(2y + 3) - 4y = 6$ which reduces to
$4y + 6 - 4y = 6$ or $6 = 6$ which is a true equation. Therefore the equations are dependent and any solution to one of the equations is a solution to both. So there are an infinite number of solutions given by any solution to $-x + 2y = -3$.

17. If $y = -x$ and $x - 3y = -12$ then substitute $-x$ for y in the 2nd equation. This gives
$x - 3(-x) = -12 \Rightarrow x + 3x = -12 \Rightarrow 4x = -12 \Rightarrow x = -3$ and $y = -(-3) = 3$.
Therefore the solution is $(-3, 3)$.

21. The 1st equation $x = 2$ gives us half the solution, namely $x = 2$. Then from the 2nd equation we see that $4 \cdot 2 - 3y = -1 \Rightarrow 8 - 3y = -1 \Rightarrow 9 = 3y \Rightarrow y = 3$.
Therefore the solution is $(2, 3)$.

Exercises 7.2

25 Since $y = x$, substitute x for y in the 2nd equation. This gives $4x + 5x = 7$ or $9x = 7$ which implies $x = \frac{7}{9}$. Since $y = x$, $y = \frac{7}{9}$. Therefore the solution is $(\frac{7}{9}, \frac{7}{9})$.

29 From eq 1 we see that $y = -3x$. Then substitute $y = -3x$ in the 2nd equation $5y = 4x$. This gives $5(-3x) = 4x \Rightarrow -15x = 4x \Rightarrow 0 = 19x \Rightarrow x = 0$ and $y = -3 \cdot 0 = 0$. Therefore the solution is $(0, 0)$.

33 From eq 2 we see that $x = 4y + 50$. Then substituting into eq 1 gives $2(4y + 50) + 3y = 1$ which simplifies to $8y + 100 + 3y = 1 \Rightarrow 11y = -99 \Rightarrow y = -9$ and $x = 4 \cdot (-9) + 50 = -36 + 50 = 14$. Therefore the solution is $(14, -9)$.

37 The system $\begin{matrix} y = 3 \\ y = -2 \end{matrix}$ is inconsistent. The lines are horizontal and do not intersect. Hence there is no solution.

41 eq 1: $2x + 2y = 14$ is equivalent to $x + y = 7 \Rightarrow y = 7 - x$.
eq 2: $3x - 3y = 9$ is equivalent to $x - y = 3 \Rightarrow x - 3 = y \Rightarrow y = x - 3$.
Therefore $7 - x = x - 3 \Rightarrow 10 = 2x \Rightarrow x = 5$.
If $x = 5$ then $y = 7 - 5 = 2$. Therefore the solution is $(5, 2)$.

45 eq 1: $3x - 2y = 14 \Rightarrow 3x - 14 = 2y \Rightarrow y = \frac{3x - 14}{2}$.
eq 2: $2x + 5y = 3$
Substitute $\frac{3x - 14}{2}$ for y in eq 2. This gives $2x + 5 \cdot \frac{3x - 14}{2} = 3$ which can be simplified by multiplying both sides by 2: $2 \cdot (2x) + 2 \cdot 5 \cdot \left(\frac{3x - 14}{2}\right) = 2 \cdot 3 \Rightarrow$
$4x + 5(3x - 14) = 6 \Rightarrow 4x + 15x - 70 = 6 \Rightarrow 19x = 76 \Rightarrow x = 4$.
Then $y = \frac{3 \cdot 4 - 14}{2} = \frac{-2}{2} = -1$ and the solution is $(4, -1)$.

49 eq 1: $2x - 5y = 5$ is equivalent to $2x - 5 = 5y$ or $y = \frac{2x - 5}{5}$.
eq 2: $-4x + 10y = 2$ after multiplying by $\frac{1}{2}$ reduces to $-2x + 5y = 1$.
Substituting for y we get
$-2x + 5 \cdot \left(\frac{2x - 5}{5}\right) = 1$ and after canceling the 5, the equation reduces to

Page 138

$-2x + 2x - 5 = 1$. This gives $-5 = 1$ which is a false statement.
Hence the equations are inconsistent and there are no solutions.

53 Line 1: $x + 4y = -4 \Rightarrow 4y = -x - 4 \Rightarrow y = -\frac{1}{4}x - 1$ so $m = -\frac{1}{4}$.
Line 2: $3x - 4y = 12 \Rightarrow 4y = 3x - 12 \Rightarrow y = \frac{3}{4}x - 3$ so $m = \frac{3}{4}$.
Line 3: $4x + 3y = -24 \Rightarrow 3y = -4x - 24 \Rightarrow y = -\frac{4}{3}x - 8$ and $m = -\frac{4}{3}$.
Since $\frac{3}{4}$ and $-\frac{4}{3}$ are negative reciprocals, the vertices form a right triangle.
To find where line 1 intersects line 2, substitute $-\frac{1}{4}x - 1$ for y in the equation for line 2.
This gives $-\frac{1}{4}x - 1 = \frac{3}{4}x - 3$ which can be simplified by multiplying both sides by 4. Then
$-x - 4 = 3x - 12 \Rightarrow 8 = 4x \Rightarrow x = 2$. Then $y = -\frac{1}{4}(2) - 1 = -\frac{1}{2} - 1 = -\frac{3}{2}$.
So one vertex is $(2, -\frac{3}{2})$.
To find where line 1 and line 3 intersect solve $-\frac{1}{4}x - 1 = -\frac{4}{3}x - 8$ which after multiplying
both sides by 12 gives $-3x - 12 = -16x - 96 \Rightarrow 13x = -84 \Rightarrow x = -\frac{84}{13}$.
Then $y = -\frac{1}{4}\left(-\frac{84}{13}\right) - 1 = \frac{21}{13} - 1 = \frac{21}{13} - \frac{13}{13} = \frac{8}{13}$. So a 2nd vertex is $(-\frac{84}{13}, \frac{8}{13})$.

The 3rd vertex is where line 2 and line 3 intersect. To find this point solve
$\frac{3}{4}x - 3 = -\frac{4}{3}x - 8$ which after multiplying by 12 becomes $9x - 36 = -16x - 96 \Rightarrow$
$25x = -60 \Rightarrow x = -\frac{60}{25} = -\frac{12}{5}$. Then $y = \frac{3}{4}\left(-\frac{12}{5}\right) - 3 = -\frac{36}{20} - \frac{60}{20} =$
$-\frac{96}{20} = -\frac{24}{5}$. Therefore the 3rd vertex is $(-\frac{12}{5}, -\frac{24}{5})$.

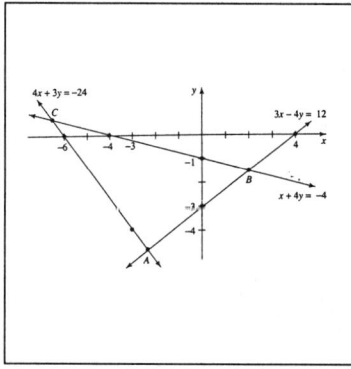

Figure 53

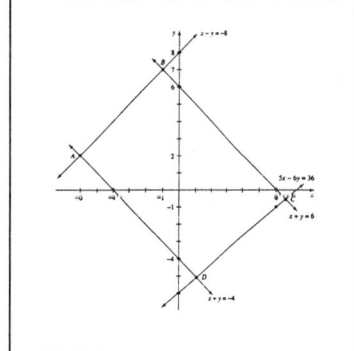

Figure 57

[57] Line 1: $x - y = -8$ is equivalent to $y = x + 8$ so $m = 1$.
Line 2: $x + y = -4$ is equivalent to $y = -x - 4$ so $m = -1$.
Line 3: $x + y = 6$ is equivalent to $y = -x + 6$ so $m = -1$.
Line 4: $5x - 6y = 36$ is equivalent to $5x - 36 = 6y \Rightarrow y = \frac{5}{6}x - 6$ so $m = \frac{5}{6}$.
From the slopes we see that lines 2 and 3 are parallel, but lines 1 and 4 are not parallel.
Therefore the figure is not a parallelogram and it can not be a rectangle.
Vertex A is the intersection of lines 1 and 2. To find this vertex set the two y values equal to each other. This gives $x + 8 = -x - 4 \Rightarrow 2x = -12 \Rightarrow x = -6$ and $y = -6 + 8 = 2$.
Therefore vertex A is $(-6, 2)$.
Vertex B is the intersection of lines 1 and 3. To find this vertex solve $x + 8 = -x + 6 \Rightarrow 2x = -2 \Rightarrow x = -1$ and $y = -x + 8 = -1 + 8 = 7$. So B is $(-1, 7)$.
Vertex C is the intersection of lines 2 and 4. Therefore solve the equation $-x - 4 = \frac{5}{6}x - 6$.
$-x - 4 = \frac{5}{6}x - 6 \Rightarrow -x = \frac{5}{6}x - 2 \Rightarrow -6x = 5x - 12 \Rightarrow 12 = 11x \Rightarrow x = \frac{12}{11}$ and
$y = -x - 4 = -\frac{12}{11} - 4 = -\frac{12}{11} - \frac{44}{11} = -\frac{56}{11}$. Therefore C is $(\frac{12}{11}, -\frac{56}{11})$.
Vertex D is the intersection of lines 3 and 4; so solve $-x + 6 = \frac{5}{6}x - 6$.
$-x + 6 = \frac{5}{6}x - 6 \Rightarrow 12 = \frac{5}{6}x + x \Rightarrow 72 = 5x + 6x \Rightarrow 72 = 11x \Rightarrow x = \frac{72}{11}$
and $y = -x + 6 = -\frac{72}{11} + 6 = -\frac{72}{11} + \frac{66}{11} = -\frac{6}{11}$. Thus the point D is $(\frac{72}{11}, -\frac{6}{11})$.

Review Problems

[61] $-4x + 6y$
       $\underline{7x - 6y}$
       $3x$

[65] $6x - 8y$
       $\underline{-6x + 8y}$
       0

Exercises 7.3

1. To solve the system $\begin{array}{c} 6x + 2y = 1 \\ 4x - 5y = -1 \end{array}$ for y multiply the top equation by -2 and the bottom equation by 3 and add. This will result in the x terms canceling out.

$$\begin{array}{r} -12x - 4y = -2 \\ 12x - 15y = -3 \\ \hline -19y = -5 \end{array}$$

which implies that $y = \dfrac{5}{19}$.

5. eq 1: $x - 2y = -5$

eq 2: $3x + 2y = 17$

a) <u>Substitution</u> <u>Method</u>

From eq 1 we see that $x = 2y - 5$. Substitute this into eq 2:

$3 \cdot (2y - 5) + 2y = 17 \Rightarrow 6y - 15 + 2y = 17 \Rightarrow 8y = 32 \Rightarrow y = 4$.

Then $x = 2 \times 4 - 5 = 8 - 5 = 3$. Therefore the solution is $(3, 4)$.

b) <u>Addition</u> <u>Method</u>

Add -3 times eq 1 to eq 2.

$$\begin{array}{r} -3x + 6y = 15 \\ 3x + 2y = 17 \\ \hline 8y = 32 \end{array}$$

which implies that $y = 4$. Substituting $y = 4$ into eq 2 gives

$3x + 2 \cdot 4 = 17 \Rightarrow 3x + 8 = 17 \Rightarrow 3x = 9 \Rightarrow x = 3$.

9. eq 1: $x = 3$

eq 2: $x - 4y = 7$

a) <u>Substitution method</u>

Substitute $x = 3$ into eq 2. Then $3 - 4y = 7 \Rightarrow -4 = 4y \Rightarrow y = -1$.

Therefore the solution is $(3, -1)$.

b) <u>Addition Method</u>

Add -1 times eq 1 to eq 2.

$$\begin{array}{r} -x = -3 \\ x - 4y = 7 \\ \hline -4y = 4 \end{array}$$

which implies that $y = -1$. Therefore the solution is $(3, -1)$.

Exercises 7.3

[13] eq 1: $3x - 2y = 10$ Add the 2 equations together to eliminate the y terms:
  eq 2: $5x + 2y = 22$
  $\overline{8x = 32}$ which means $x = 4$. Then replace x by 4 in eq 2:

$$5 \cdot 4 + 2y = 22 \Rightarrow 20 + 2y = 22 \Rightarrow 2y = 2 \Rightarrow y = 1.$$

The solution is $(4, 1)$.

[17] eq 1: $4x - 3y = 7$ Multiply eq 1 by 5 and eq 2 by 3. Then add to eliminate the y term.
  eq 2: $6x + 5y = 1$

$$\begin{array}{r} 20x - 15y = 35 \\ 18x + 15y = 3 \\ \hline 38x = 38 \end{array}$$

which implies that $x = 1$. Replacing x by 1 in eq 1 gives $4 \cdot 1 - 3y = 7 \Rightarrow 4 - 3y = 7 \Rightarrow -3 = 3y \Rightarrow y = -1$ and the solution is $(1, -1)$.

[21] eq 1: $3x - 5y = 0$ Multiply eq 1 by -2 and add to eq 2.
  eq 2: $6x - 10y = 0$

$$\begin{array}{r} -6x + 10y = 0 \\ 6x - 10y = 0 \\ \hline 0 = 0 \end{array}$$

This implies that the equations are dependent. The two lines are the same and any solution to one equation is a solution to the other equation. Therefore the solution is any solution to $3x - 5y = 0$.

[25] eq 1: $6x + 4y = 14$ Add $-1 \times$ eq 1 to eq 2.
  eq 2: $6x + 9y = 15$

$$\begin{array}{r} -6x - 4y = -14 \\ 6x + 9y = 15 \\ \hline 5y = 1 \end{array}$$

which implies that $y = \frac{1}{5}$. Substitute $y = \frac{1}{5}$ into equation 1: $6x + 4 \cdot \frac{1}{5} = 14 \Rightarrow 6x + \frac{4}{5} = \frac{70}{5} \Rightarrow 6x = \frac{66}{5} \Rightarrow x = \frac{66}{5} \cdot \frac{1}{6} = \frac{66}{30} = \frac{6 \times 11}{6 \times 5} = \frac{11}{5}$. Therefore the solution is $(\frac{11}{5}, \frac{1}{5})$.

Exercises 7.3

29 eq 1: $3x + 2y = 19$ Multiply eq 1 by -5 and multiply eq 2 by 2. Then add the

 eq 2: $4x + 5y = 37$ equations to eliminate y.

$$\begin{array}{r} -15x - 10y = -95 \\ 8x + 10y = 74 \\ \hline -7x = -21 \end{array}$$ which implies that $x = \dfrac{-21}{-7} = 3$.

Then $3 \cdot 3 + 2y = 19 \Rightarrow 2y = 19 - 9 \Rightarrow 2y = 10 \Rightarrow y = 5$.

Therefore the solution is $(3, 5)$.

33 eq 1: $9x - 4y = 0$ Substitute $3y$ for x in eq 1. This gives

 eq 2: $x = 3y$ $9 \cdot (3y) - 4y = 0 \Rightarrow 27y - 4y = 0 \Rightarrow 23y = 0 \Rightarrow y = 0$.

 If $y = 0$ then $x = 3 \cdot 0 = 0$ and the solution is $(0, 0)$.

37 eq 1: $-2x + 3y = 5$ Multiply eq 1 by 2 and add to eq 2:

 eq 2: $4x - 6y = 5$

$$\begin{array}{r} -4x + 6y = 10 \\ 4x - 6y = 5 \\ \hline 0 = 15 \end{array}$$ which implies that the equations are inconsistent

and there are no solutions.

41 eq 1: $7x - 4y = 56$ Multiply eq 1 by -3 and eq 2 by 7. This gives

 eq 2: $3x - 5y = -45$

$$\begin{array}{r} -21x + 12y = -168 \\ 21x - 35y = -315 \\ \hline -23y = -483 \end{array}$$ which implies that $y = \dfrac{-483}{-23} = 21$.

and $3x - 5 \cdot 21 = -45 \Rightarrow 3x - 105 = -45 \Rightarrow 3x = 60 \Rightarrow x = 20$.

Therefore the solution is $(20, 21)$.

45 eq 1: $x = -3$ We know $x = -3$ so replace x by -3 in eq 2 to find y:

 eq 2: $2x - 3y = -1$ $2 \cdot (-3) - 3y = -1 \Rightarrow -6 - 3y = -1 \Rightarrow -3y = 5 \Rightarrow$

$y = -\dfrac{5}{3}$. Therefore the solution is $\left(-3, -\dfrac{5}{3}\right)$.

Exercises 7.3

49 eq 1: $y = \frac{1}{2}x + 2$ Replace $\frac{1}{2}x + 2$ for y in eq 2: $\frac{1}{2}x + 2 = \frac{1}{3}x - 2$.

eq 2: $y = \frac{1}{3}x - 2$ Simplify by multiplying both sides by 6.

$$6\left(\tfrac{1}{2}x + 2\right) = 6\left(\tfrac{1}{3}x - 2\right) \;\Rightarrow\; 3x + 12 = 2x - 12 \;\Rightarrow\; x = -24$$

and $y = \frac{1}{2} \cdot (-24) + 2 = -12 + 2 = -10$. So the solution is $(-24, -10)$.

53 eq 1: $\frac{x}{2} + \frac{y}{3} = 2$ when multiplied by 6 simplifies to $3x + 2y = 12$.

eq 2: $\frac{x}{4} - \frac{y}{6} = 1$ when multiplied by 12 simplifies to $3x - 2y = 12$.

                    Add the 2 equations: $\overline{6x \;=\; 24}$ which means $x = 4$.

Substitute $x = 4$ into $3x + 2y = 12$. This gives $12 + 2y = 12 \;\Rightarrow\; 2y = 0 \;\Rightarrow\; y = 0$.
Therefore the solution is $(4, 0)$.

57 eq 1: $y = x + 3000$ Replace y by $x + 3000$ in eq 2. This gives

eq 2: $.10x + .08y = 600$ $.10x + .08(x + 3000) = 600 \;\Rightarrow\; .10x + .08x + 240 = 600 \;\Rightarrow$

$$.18x = 360 \;\Rightarrow\; x = \frac{360}{.18} = 2000 \text{ and}$$

$y = x + 3000 = 2000 + 3000 = 5000$. So the solution is $(2000, 5000)$.

61 eq 1: $.3x + .2y = 1.4$ is equivalent to $3x + 2y = 14$.

eq 2: $.7x - y = .02$ implies that $y = .7x - .02$.

Replace y by $.7x - .02$ in the equation $3x + 2y = 14$. Then we see that

$$3x + 2(.7x - .02) = 14 \;\Rightarrow\; 3x + 1.4x - .04 = 14 \;\Rightarrow\; 4.4x = 14.04 \;\Rightarrow\; x = \frac{14.04}{4.4} \approx 3.191.$$

$y = .7x - .02 \approx .7 \cdot (3.191) - .02 \approx 2.214$.

65 eq 1: $y = \frac{2}{3}x - 1$

eq 2: $\frac{x}{7} + \frac{y+1}{2} = \frac{5}{14}$ which when multiplied by 14 reduces to $2x + 7(y+1) = 5 \;\Rightarrow$

$2x + 7y = -2$. Substitute $\frac{2}{3}x - 1$ for y in this equation. This gives

$$2x + 7\left(\tfrac{2}{3}x - 1\right) = -2 \;\Rightarrow\; 2x + \tfrac{14}{3}x - 7 = -2 \;\Rightarrow\; \tfrac{20}{3}x = 5 \;\Rightarrow\; x = 5 \cdot \tfrac{3}{20} = \tfrac{15}{20} = \tfrac{3}{4}.$$

$y = \frac{2}{3} \cdot \left(\frac{3}{4}\right) - 1 = \frac{1}{2} - 1 = -\frac{1}{2}$. Therefore the solution is $\left(\frac{3}{4}, -\frac{1}{2}\right)$.

69 To solve the system $\frac{1}{x} + \frac{1}{y} = 3$ let $a = \frac{1}{x}$ and $b = \frac{1}{y}$. Then, in terms of a and b, we get
$\frac{1}{x} - \frac{1}{y} = 5$

$\begin{array}{l} a + b = 3 \\ a - b = 5 \\ \hline 2a = 8 \end{array}$ Solve this system by adding the 2 equations to eliminate b.

$\Rightarrow a = 4$.

Then $4 + b = 3 \Rightarrow b = -1$. If $a = 4$ then $\frac{1}{x} = 4$ implying that $x = \frac{1}{4}$.

If $b = -1$ then $\frac{1}{y} = -1$ implying that $y = -1$.

Therefore the solution to the original system is $(\frac{1}{4}, -1)$.

Review Problems

77 The graph of $2x - 3y < 6$ is a half-plane bounded by, but not including the line $2x - 3y = 6$.

So graph $2x - 3y = 6$ as a dotted line and use $(0, 0)$ as a test point to determine which half-plane is correct.

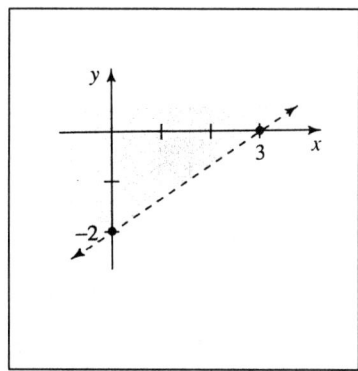

Page 145

Exercises 7.4

[1] To graph $y \leq x+2$ first graph the line $y = x+2$. Note that $(-2, 0)$, $(0, 2)$, and $(1, 3)$ are on the line. Use a solid line since the line is part of the graph. Then test $(0, 0)$ in the inequality $y \leq x+2$. It satisfies the inequality implying that the half-plane includes the origin $(0, 0)$. So shade in the region below the line.

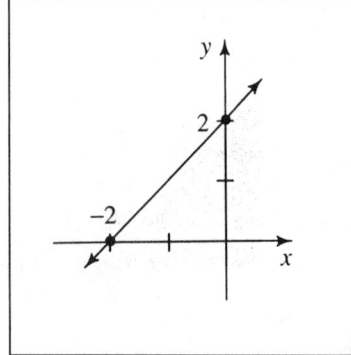

Figure 1

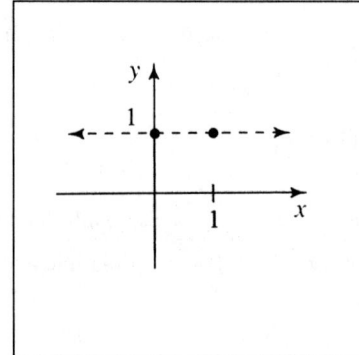

Figure 5

[5] $y < 1$ is the half bounded by, but not including the horizontal line $y = 1$. The half-plane opens downward because the inequality is $<$.

[9] To graph the solution set of the system $\begin{array}{c} x \geq 1 \\ y < 1 \end{array}$ graph each inequality separately and find the intersection of the 2 half-planes. $x \geq 1$ is the half-plane opening to the right of the vertical line $x = 1$, including the line. The half-plane $y < 1$ is the half-plane opening below the horizontal line $y = 1$, not including the line. The intersection is indicated in the graph below.

Exercises 7.4

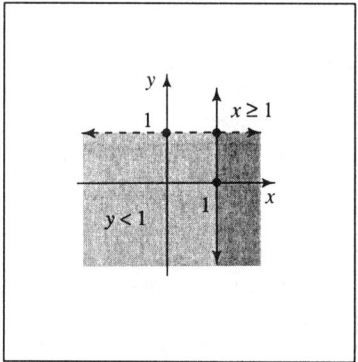

Figure 9

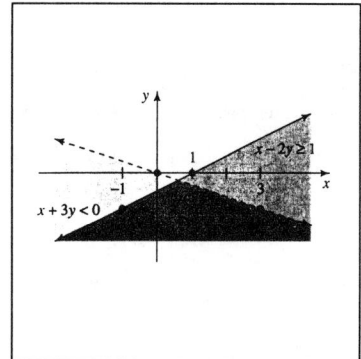

Figure 13

13 To graph the solution set of $\begin{array}{c} x + 3y < 0 \\ x - 2y \geq 1 \end{array}$ first graph $x + 3y = 0$ as a dotted line. Test $(0, 1)$ in the inequality $x + 3y < 0$. It does not satisfy the inequality implying that the half-plane opens downward because $(0, 1)$ is not part of the graph. Then graph the line $x - 2y = 1$ as a solid line. Testing $(0, 0)$ we see it does not satisfy the inequality implying that this half-plane also opens downward. The solution set is the intersection of the two half-planes.

17 Graph $2x + 3y = 6$ as a solid line and shade the region below the line because $(0, 0)$ satisfies the inequality $2x + 3y \leq 6$. Then graph $2x - 3y = 6$ as as a dotted line and shade the region above the line because the half-plane must include the point $(0, 0)$. The solution set is given by the intersection of the half-planes.

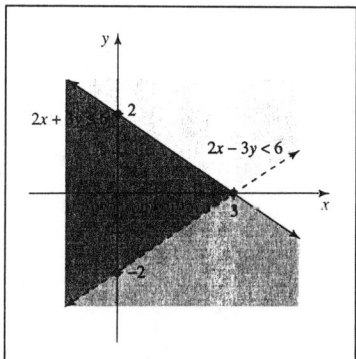

Figure 17

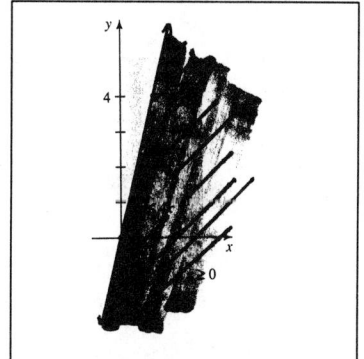

Figure 21

Exercises 7.4

21 The half-plane $x \geq 0$ is the half-plane bounded by, and including the y-axis ($x = 0$) opening to the right. The half-plane $y \leq 4x$ is the half-plane bounded by, and including the line $y = 4x$ opening to the right. To see this, test $(1, 0)$ in $y \leq 4x$ which gives $0 \leq 4$, a true statement. Hence the half-plane must include this point.

25 To graph $\begin{array}{c} x + 3y < 1 \\ x - y \leq 2 \end{array}$ first graph $x + 3y = 1$ as a dotted line and shade the region below.
Then graph $x - y = 2$ as a solid line and shade the region above. The solution is set is given by the intersection of the two graphs.

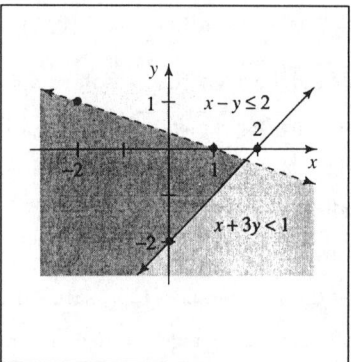

Figure 25

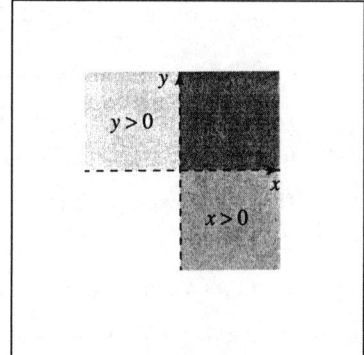

Figure 29

29 The graph of $\begin{array}{c} x > 0 \\ y > 0 \end{array}$ consists of all the points in the xy-plane which have positive x and y coordinates. Therefore the graph is the set of points in the 1st quadrant.

33 To graph $\begin{array}{c} \frac{x-1}{15} + \frac{2y}{5} < \frac{1}{3} \\ \frac{y+1}{4} > \frac{x}{12} \end{array}$ first simplify the inequalities.

$\frac{x-1}{15} + \frac{2y}{5} < \frac{1}{3} \Rightarrow 15 \cdot \left(\frac{x-1}{15} + \frac{2y}{5}\right) < 15 \cdot \frac{1}{3} \Rightarrow x - 1 + 6y < 5 \Rightarrow x + 6y < 6$.

$\frac{y+1}{4} > \frac{x}{12} \Rightarrow 12 \cdot \frac{y+1}{4} > 12 \cdot \frac{x}{12} \Rightarrow 3y + 3 > x \Rightarrow 3 > x - 3y \Rightarrow x - 3y < 3$.

Then graph the line $x + 6y = 6$ as a dotted line and shade the half-plane below the line.
Graph the line $x - 3y = 3$ as a dotted line and shade the half-plane above the line. Note that

Exercises 7.4

the half-plane opens above the line despite the inequality being $<$. Use test points. The intersection is the graph of the solution set.

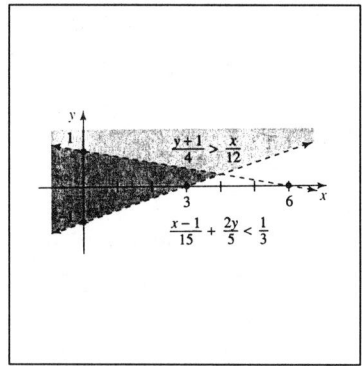

Figure 33

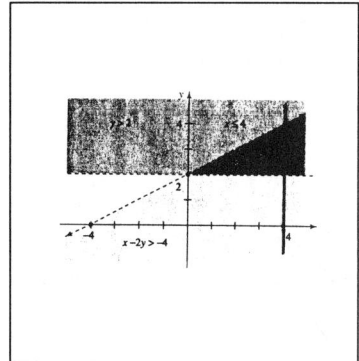

Figure 37

[37] The graph of $x - 2y > -4$ is the intersection of the 3 half-planes.
$x \leq 4$
$y > 2$

The half-plane $x - 2y > -4$ is the half-plane bounded by, but not including the line $x - 2y = -4$. This half-plane opens downward because (0, 0) satisfies $x - 2y > -4$. Note that $x - 2y > -4$ opens downward despite the "greater than" sign.
$x \leq 4$ is the half opening to the left bounded by the vertical line $x = 4$ and $y > 2$ is the half-plane opening upward bounded by the horizontal line $y = 2$. Draw $x - 2y = -4$ and $y = 2$ with dotted lines and draw $x = 4$ with a solid line.

Review Problems

[45] Let w be the width and $3w - 2$ the length. Since the perimeter is 44 m,
$2w + 2(3w - 2) = 44 \Rightarrow 2w + 6w - 4 = 44 \Rightarrow 8w = 48 \Rightarrow w = 6$.
Therefore the width is 6 m and the length is $3 \times 6 - 2 = 16$ m .

Exercises 7.5

1. Let x and y be the two numbers. Their sum being 38 implies that $x + y = 38$ and their difference being 8 implies that $x - y = 8$. To eliminate y, add the two equations together:

$$x + y = 38$$
$$x - y = 8$$
$$\overline{2x \ \ = 46}$$

Therefore $x = 23$ and $23 - y = 8$ means $y = 15$.

5. Let x and y be the numbers. If one number exceeds the other by 7 then, assuming y is the larger, $y = x + 7$. If 3 times the larger number plus 2 times the smaller number is 8 then $3y + 2x = 8$. Substitute $x + 7$ for y in the 2nd equation. This gives

$$3(x+7) + 2x = 8 \Rightarrow 3x + 21 + 2x = 8 \Rightarrow 5x = -13 \Rightarrow x = -\tfrac{13}{5}.$$

Then $y = -\tfrac{13}{5} + 7 = -\tfrac{13}{5} + \tfrac{35}{5} = \tfrac{22}{5}$. Therefore the two numbers are $-\tfrac{13}{5}$ and $\tfrac{22}{5}$.

9. Let x be the numerator and y the denominator. Then $x + y = 30$ and $\tfrac{x+3}{y+3} = \tfrac{4}{5}$. Simplify the 2nd equation by cross-multiplying: $5x + 15 = 4y + 12 \Rightarrow 5x - 4y = -3$.

Multiply the 1st equation by 4:
$$4x + 4y = 120$$
$$5x - 4y = -3$$
$$\overline{9x \ \ = 117}$$

$\Rightarrow x = \tfrac{117}{9} = 13$. Then $13 + y = 30 \Rightarrow y = 17$. Therefore the original fraction is $\tfrac{13}{17}$.

13. Problem Statement: Pylades has 28 dimes and quarters worth $5.20. How many of each type of coin does he have?

Let x be the number of dimes and y the number of quarters. Then $x + y = 28$ and the value of the dimes is $10x$ cents and the value of the quarters is $25y$ cents. Therefore

$$10x + 25y = 520 \quad \Leftarrow \text{ Add this to } -10 \times \text{1st equation.}$$
$$\underline{-10x - 10y = -280}$$
$$15y = 240$$

$\Rightarrow y = \tfrac{240}{15} = 16$ and $x + 16 = 28 \Rightarrow x = 12$.

Hence there are 12 dimes and 16 quarters.

21. Problem Statement: Avi invested some money in bonds at 6% and an additional $800 in stocks at 7%. The amount of interest earned in a year at 7% was $4 less than twice the amount earned at 6%. How much did he invest at each rate?

Exercises 7.5

Let x be the amount invested at 6% and y the amount invested at 7%.
Then $y = x + 800$ and $.07y = 2(.06x) - 4$. If we multiply the 2nd equation by 100 we get
$7y = 12x - 400$. Substituting $x + 800$ for y gives $7(x + 800) = 12x - 400 \Rightarrow$
$7x + 5600 = 12x - 400 \Rightarrow 6000 = 5x \Rightarrow x = \frac{6000}{5} = 1200$ and $y = 2000$.
So $1200 was invested at 6% and $2000 was invested at 7%.

25 Problem Statement: Mr. Wilson has 20 pounds of candy worth 80¢/lb. He is going to mix the candy with nuts worth 50¢/lb to get a mixture worth 60¢/lb. How many pounds of nuts are needed and how many pounds are in the mixture?

Let x be the number of pounds of nuts needed in the mixture and let y be the total number of pounds in the mixture. Then $y = 20 + x$ and if the mixture is worth 60¢/lb, the mixture should cost a total of $.60y$. But we know this mixture costs $20(.80) + x(.50)$ or $16 + .5x$. Therefore $16 + .5x = .6y$ or $-.5x + .6y = 16$. Substitute $20 + x$ for y in this equation. This gives
$-.5x + .6(20 + x) = 16 \Rightarrow -.5x + 12 + .6x = 16 \Rightarrow .1x = 4 \Rightarrow x = \frac{4}{.1} = 40$.
Therefore he needs 40 pounds of nuts and there is a total of 60 pounds in the mixture.

29 eq 1: $x + y = 3$
eq 2: $.06x + .15y = .12(3) \Rightarrow 100(.06x + .15y) = 100(.36) \Rightarrow 6x + 15y = 36$.
Multiply eq 1 by -6 and add to eq 2: $-6x - 6y = -18$
$ 6x + 15y = 36$
$ 9y = 18 \Rightarrow y = 2$ and $x + 2 = 3 \Rightarrow x = 1$

Therefore the solution is $x = 1$ and $y = 2$.

33 Problem Statement: A farmer has 70 gallons of a 60% disinfectant. How much water does she have to add to it to obtain a 50% solution? How many gallons are in the mixture?

Let x be the amount of water that is added and let y be the total amount in the mixture.
Then $y = 70 + x$. Since only water was added the amount of disinfectant in the mixture is
$.60(70) = 42$ gallons. If the mixture must be 50% then $.50y = 42 \Rightarrow y = \frac{42}{.5} = 84$.
Then $84 = 70 + x \Rightarrow x = 14$. Therefore the farmer must add 14 gallons of water, and the total mixture is 84 gallons.

Exercises 7.5

37 Let l be the length and w the width. Since the width is 2 m less than half its length, $w = \frac{1}{2}l - 2$. The perimeter is 56 m implies that $2l + 2w = 56$. Substituting for w in the 2nd equation gives $2l + 2\left(\frac{1}{2}l - 2\right) = 56 \Rightarrow 2l + l - 4 = 56 \Rightarrow 3l = 60 \Rightarrow l = 20$ and $w = \frac{1}{2} \cdot 20 - 2 = 10 - 2 = 8$. Therefore the length is 20 m and the width is 8 m.

41 Problem Statement: Two angles are supplementary and the larger angle is three times the smaller angle. Find the angles.

Let x and y be the angles. When angles are supplementary they add up to 180°. Therefore $x + y = 180$ and if y is the larger angle, $y = 3x$. Substituting $3x$ for y in the 1st equation: $x + 3x = 180 \Rightarrow 4x = 180 \Rightarrow x = 45$. Thus the two angles are 45° and 135°.

45 Let x and y be the other two angles, x the larger of the two. Then $x - y = 10$ and since the sum of all 3 angles must be 180°, $50 + x + y = 180$ or $x + y = 130$.

Adding the two equations gives

$x - y = 10$
$x + y = 130$
$\overline{2x = 140} \Rightarrow x = 70$ and $y = 60$. So the other two angles are 70° and 60°.

49 Problem Statement: For two hours Adolfo flies 300 miles with the wind and for 5 hours he flies 350 miles against the wind. Find the speed of the plane in still air and the speed of the wind.

Let x be the speed of the plane in still air and y the speed of the wind. With the wind he has a rate of $x + y$ and against the wind he has a rate of $x - y$. Therefore, since he flies 300 miles with the wind and $D = RT$,

$2(x + y) = 300 \Rightarrow x + y = 150$. Similarly, since he flies 350 miles against the wind,
$5(x - y) = 350 \Rightarrow x - y = 70$. Add the two equations together to eliminate y:
$\overline{ 2x = 220} \Rightarrow x = 110$ and $110 + y = 150 \Rightarrow y = 40$.

Therefore the speed of the plane in still air is 110 mph and the wind speed is 40 mph.

53 Problem Statement: If a boat can travel 22 mph with the current and 11 mph against the current, find the speed of the boat in still water and the rate of the current.

Let x be the speed of the boat in still water and y the rate of the current. Then

Exercises 7.5

$x + y = 22$ and
$\underline{x - y = 11}$
$2x = 33$ $\Rightarrow$ $x = 16.5$ and $16.5 + y = 22$ $\Rightarrow$ $y = 22 - 16.5 = 5.5$.

Therefore the speed of the boat in still water is 16.5 mph and the rate of the current is 5.5 mph .

[57] Problem Statement: Aurora's sister is 1 year older than twice Aurora's age. In 6 years the sum of their ages will be 55 . Find their present ages.

Let x be Aurora's age and y her sister's age. Then

$y = 2x + 1$

$(x + 6) + (y + 6) = 55$ $\Rightarrow$ $x + y = 43$. Substituting $2x + 1$ for y in the last equation gives

$x + 2x + 1 = 43$ $\Rightarrow$ $3x = 42$ $\Rightarrow$ $x = 14$ and $y = 28 + 1 = 29$.

Therefore Aurora's age is 14 and her sister's age is 29 .

[61] Problem Statement: Two doctors perform an operation and share the fee of $12,000 in the ratio of 11 : 13 . How much money does each doctor receive?

Let x and y be their fees, x being the smaller one. Then

$x + y = 12{,}000$

$\frac{x}{y} = \frac{11}{13}$ $\Rightarrow$ $13x = 11y$ $\Rightarrow$ $13x - 11y = 0$. Multiply the 1st equation by 11 and add to the last equation. $11x + 11y = 132{,}000$
$\underline{13x - 11y = 0}$
$24x = 132{,}000$ $\Rightarrow$ $x = \frac{132{,}000}{24} = 5{,}500$ and

$5{,}500 + y = 12{,}000$ $\Rightarrow$ $y = 12{,}000 - 5{,}500 = 6{,}500$. Therefore the two fees are

$5,500 and $6,500 .

[65] Problem Statement: A carpenter has cut a 25 foot board into two pieces in such a way that one piece is 1 foot longer than twice the other piece. How long is each piece?

Let x and y be the two pieces. Then $x + y = 25$ and $y = 2x + 1$. Substituting $2x + 1$ for y in the 1st equation gives $x + 2x + 1 = 25$ $\Rightarrow$ $3x = 24$ $\Rightarrow$ $x = 8$ and $y = 16 + 1 = 17$.

Therefore the pieces are 8 ft and 17 ft.

[69] Problem Statement: The sum of the lengths of the Suez Canal and the Panama Canal is 250 km. If the Suez Canal were 4 km shorter, then it would be twice the length of the Panama Canal. Find the length of each canal.

Let x be the length of the Suez Canal and y the length of the Panama Canal. Then
$x + y = 250$ and $x - 4 = 2y \Rightarrow x = 2y + 4$. Substituting for x in the 1st equation gives
$2y + 4 + y = 250 \Rightarrow 3y = 246 \Rightarrow y = 82$ and $x = 2 \cdot 82 + 4 = 164 + 4 = 168$.
Therefore the Suez Canal is 168 km and the Panama Canal is 82 km.

[73] Problem Statement: Antonio leaves Florence at noon and drives south toward Rome. An hour later Lorenzo follows him, and at 4 PM Antonio is 145 km ahead of Lorenzo. If Lorenzo had driven north, then at 5 PM they would have been 840 km apart. How fast do they each drive?

Let x be Antonio's speed and y Lorenzo's speed. At 4 PM Antonio has driven 4 hours and Lorenzo has driven 3 hours. Since $D = RT$ and the difference of their distances is 145 we have $4x - 3y = 135$. If they travel in opposite directions then the distance apart is the sum of their distances. At 5 PM Antonio would have driven 5 hours and Lorezo would have driven 4 hours. Therefore $5x + 4y = 840$.

$4x - 3y = 145 \Rightarrow 16x - 12y = 580$ (after multiplying by 4).
$5x + 4y = 840 \Rightarrow 15x + 12y = 2520$ (after multiplying by 3).
$\overline{31x = 3100} \Rightarrow x = 100$ and $400 - 3y = 145 \Rightarrow 3y = 255 \Rightarrow$
$y = \frac{255}{3} = 85$. Therefore Antonio's speed is 100 kph and Lorenzo's speed is 85 kph.

[77] Problem Statement: The second side of a triangle is 4 m more than the first side. The third side is 2 m less than the sum of the first two sides. If the perimeter of the triangle is 38 m, find the 3 sides.

Let x be the 1st side and y the 2nd side. Then $x + y - 2$ is the length of the 3rd side.
Since the 2nd side is 4 m more than the 1st side, $y = x + 4$.
Since the perimeter is 38, $x + y + x + y - 2 = 38 \Rightarrow 2x + 2y = 40 \Rightarrow x + y = 20$.
Substituting $x + 4$ for y in the last equation gives $x + x + 4 = 20 \Rightarrow 2x = 16 \Rightarrow x = 8$
and $y = 6 + 4 = 12$. The 3rd side is $8 + 12 - 2 = 18$.
Therefore the 3 sides are 8 m, 12 m, and 18 m.

Chapter Review Exercises

5. $x + y = 3$
    $x - y = 1$

    The solution is $(2,1)$.

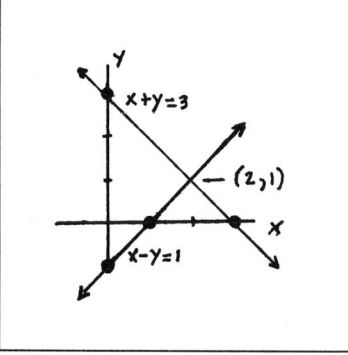

9. eq 1: $3x - y = 1$ ⇒ $y = 3x - 1$. Substitute for y in equation 2.
    eq 2: $5x - 2y = 3$

    $5x - 2(3x - 1) = 3$ ⇒ $5x - 6x + 2 = 3$ ⇒ $-x = 1$ ⇒ $x = -1$.
    Then $y = 3(-1) - 1 = -4$. Therefore the solution is $(-1, -4)$.

13. eq 1: $10x - 15y = 5$

    eq 2: $4x - 6y = 2$ ⇒ $2x - 3y = 1$ (after dividing each term by 2).

    Add eq 1 to $-5 \times$ eq 2:

    $$\begin{aligned} 10x - 15y &= 5 \\ -10x + 15y &= -5 \\ \hline 0 &= 0 \end{aligned}$$ This implies that the equations are dependent
    and there are infinitely many solutions given by $4x - 6y = 2$ or more simply $2x - 3y = 1$.

17. eq 1: $3x + 2y = -2$

    eq 2: $4y = -6x + 8$ ⇒ $6x + 4y = 8$. Add $-2 \times$ eq 1 to eq 2:

    $$\begin{aligned} -6x - 4y &= 4 \\ 6x + 4y &= 8 \\ \hline 0 &= 12 \end{aligned}$$

    This implies the equations are inconsistent and there are no solutions.

Chapter 7 Review Exercises and Test

21. eq 1: $x - 4y = 2$

    eq 2: $3x + 2y = 1$. Add $-3 \times$ eq 1 to eq 2:

    $$\begin{array}{r} -3x + 12y = -6 \\ 3x + 2y = 1 \\ \hline 14y = -5 \end{array} \Rightarrow y = -\frac{5}{14}. \text{ Then } x - 4\left(-\frac{5}{14}\right) = 2 \Rightarrow$$

    $x + \frac{20}{14} = \frac{14}{7} \Rightarrow x = \frac{14}{7} - \frac{10}{7} = \frac{4}{7}$. Therefore the solution is $(\frac{4}{7}, -\frac{5}{14})$.

25. The graph of $y \leq 3x + 4$ is the intersection of the two half-planes. $y \leq 3x + 4$ is the half-plane
    $y \geq -x - 1$
    bounded by the line $y = 3x + 4$ and $y \geq -x - 1$ is the half-plane bounded by the line
    $y = -x - 1$. Using $(0, 0)$ as a test point in each inequality, we see that the 1st half-plane opens
    downward and the 2nd opens upward. Use solid lines when graphing the two lines.

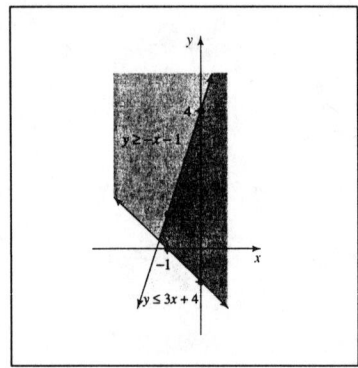

Figure 25

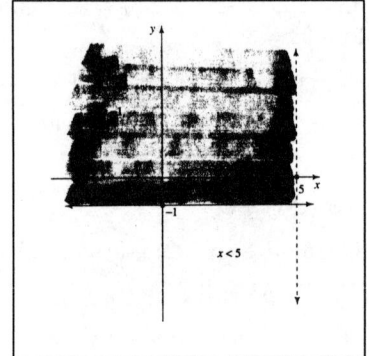

Figure 29

29. To graph the solution set of $\begin{array}{c} x < 5 \\ y \geq -1 \end{array}$ note that $x < 5$ is the half-plane opening to the left of,
    but not including the vertical line $x = 5$. $y \geq -1$ is the half-plane lying above, and including
    the horizontal line $y = -1$. The solution set is the intersection of the two half-planes.

33. Let x be the amount at 5% and y the amount at 9%. Then $x + y = 10,000 \Rightarrow y = 10,000 - x$.
    Also $.05x + 60 = .09y$. Substitute $10,000 - x$ for y in this equation which gives
    $.05x + 60 = .09(10,000 - x) \Rightarrow .05x + 60 = 900 - .09x \Rightarrow .14x = 840 \Rightarrow$
    $x = \frac{840}{.14} = 6000$. Therefore $6000 is invested at 5% and $4000 at 9%.

Chapter 7 Review Exercises and Test

37. Let x be the speed of the boat in still water and y the speed of the river. Then
$2.5(x+y) = 30 \Rightarrow x+y = 12$ (after dividing each term by 2.5).
$4.2(x-y) = 21 \Rightarrow x-y = 5$ (after dividing each term by 4.2).
Adding the two equations gives $2x = 17 \Rightarrow x = 8.5$ and $y = 12 - 8.5 = 3.5$.
So the boat travels 8.5 mph in still water and the speed of the river is 3.5 mph.

Chapter Test

1. The point $(-1, -4)$ is a solution to $y = 3x - 1$ because $-4 = 3(-1) - 1$ and
$y = -4x - 8$
$-4 = -4(-1) - 8$ are both true statements.

5. eq 1: $y = -3x + 4$
eq 2: $6x + 2y = 8 \Rightarrow 3x + y = 4$ (after dividing each term by 2). Then
$3x + (-3x + 4) = 4 \Rightarrow 4 = 4$ which means the equations are dependent and the solutions are given by $y = -3x + 4$.

9. eq 1: $y = 2x + 5$
eq 2: $x + 3y - 6$. Therefore $x + 3(2x + 5) = -6 \Rightarrow x + 6x + 15 = -6 \Rightarrow$
$7x = -21 \Rightarrow x = -3$ and $y = 2 \cdot (-3) + 5 = -1$. The solution is $(-3, -1)$.

13. eq 1: $6x - 5y = 3$
eq 2: $2y + 3x = 1 \Rightarrow 3x + 2y = 1$. Multiply eq 2 by -2 and add to eq 1:
$$\begin{array}{r} 6x - 5y = 3 \\ -6x - 4y = -2 \\ \hline -9y = 1 \end{array}$$
$\Rightarrow y = -\frac{1}{9}$ and $3x - \frac{2}{9} = 1 \Rightarrow 3x = \frac{11}{9} \Rightarrow$
$x = \frac{1}{3} \cdot \frac{11}{9} = \frac{11}{27}$. Therefore the solution is $(\frac{11}{27}, -\frac{1}{9})$.

17. The graph of $\begin{array}{l}y \leq 2x+1\\ y \leq -x-2\end{array}$ is the intersection of these two half-planes. To graph the half-planes draw $y = 2x+1$ and $y = -x-2$ as solid lines. Then note that each half-plane lies below its boundary line. To see this, test $(0, 0)$ in each inequality to determine whether or not $(0, 0)$ is included in the graph.

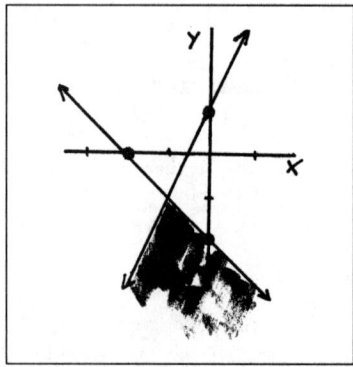

21. Let x be the number of nickels and y the number of dimes. Then
    $x + y = 40$ and $5x + 10y = 310$. Divide each term of the 2nd equation by 5 which gives
    $x + 2y = 62$. Then add -1 times the 1st equation to the 2nd equation:

    $$\begin{array}{r}-x - y = -40\\ x + 2y = 62\\ \hline y = 22\end{array}$$ and $x + 22 = 40 \Rightarrow x = 18$.

    So there are 18 nickels and 22 dimes.

25. Let x be the highest grade and y the lowest grade. Then
    $\frac{x+y}{2} = 58 \Rightarrow x + y = 116$ and $x = 5y + 14$. Therefore
    $(5y + 14) + y = 116 \Rightarrow 6y = 102 \Rightarrow y = 17$ and $x = 5 \cdot (17) + 14 = 85 + 14 = 99$.
    So the highest grade is 99 and the lowest grade is 17.

Chapter 8: Exponents and Radicals

Exercises 8.1

1. $x^{12} \cdot x^4 = x^{12+4} = x^{16}$.

5. $(-t^2)^{41} = ([-1]t^2)^{41} = (-1)^{41}(t^2)^{41} = (-1) \cdot t^{2 \cdot 41} = -t^{82}$.

9. $\left(\dfrac{k^{30}}{k^6}\right)^3 = (k^{30-6})^3 = (k^{24})^3 = k^{3 \cdot 24} = k^{72}$.

13. 0^0 is undefined.

17. $\left(\dfrac{3}{4}\right)^{-1} = \left(\dfrac{4}{3}\right)^1 = \dfrac{4}{3}$.

21. $(x^{-6})^4 = x^{-6 \cdot 4} = x^{-24} = \dfrac{1}{x^{24}}$.

25. $(x^2 y^{-3} z)^3 = (x^2)^3 \cdot (y^{-3})^3 \cdot z^3 = x^{2 \cdot 3} \cdot y^{-3 \cdot 3} \cdot z^3 = x^6 y^{-9} z^3 = x^6 \cdot \dfrac{1}{y^9} \cdot z^3 = \dfrac{x^6 z^3}{y^9}$.

29. $\left(\dfrac{a^2 b^{-1}}{c^3}\right)^{-4} = \left(\dfrac{c^3}{a^2 b^{-1}}\right)^4 = \left(\dfrac{bc^3}{a^2}\right)^4 = \dfrac{(bc^3)^4}{(a^2)^4} = \dfrac{b^4 (c^3)^4}{a^8} = \dfrac{b^4 c^{12}}{a^8}$.

33. $\dfrac{x^{10}}{x^{14}} = x^{10-14} = x^{-4}$.

37. $\dfrac{(3x^4)^{-2}}{9xx^{-6}} = \dfrac{x^6}{(3x^4)^2 \cdot 9x} = \dfrac{x^6}{3^2(x^4)^2 \cdot 9x} = \dfrac{x^6}{9x^8 \cdot 9x} = \dfrac{x^6}{81x^9} = \dfrac{1}{81x^{9-6}} = \dfrac{1}{81x^3}$.

41. It is false that $6^{-3} \cdot 6^3 = 6$; $\;6^{-3} \cdot 6^3 = 6^{-3+3} = 6^0 = 1$.

45. It is false that $(x^{-4})^{-2} = x^{-6}$ because we don't add the exponents, we multiply them.
$$(x^{-4})^{-2} = x^{(-4) \cdot (-2)} = x^8$$

Exercises 8.1

49 It is false that $(x+6)^{-2} = \frac{1}{x^2} + \frac{1}{36}$ since $(a+b)^n \neq a^n + b^n$.
$(x+6)^{-2} = \frac{1}{(x+6)^2} = \frac{1}{x^2 + 12x + 36}$.

53 $\frac{x^{-1}+y}{y^{-1}+x} = \frac{x^{-1}}{y^{-1}} + \frac{y}{x}$ is false because $\frac{a+b}{c+d} \neq \frac{a}{c} + \frac{b}{d}$.

57 $\frac{2}{a^{-1}+b^{-1}} = \frac{2}{\frac{1}{a}+\frac{1}{b}} = \frac{2 \cdot ab}{\left(\frac{1}{a}+\frac{1}{b}\right) \cdot ab} = \frac{2ab}{\frac{1}{a} \cdot ab + \frac{1}{b} \cdot ab} = \frac{2ab}{b+a}$.

61 $(2x^{-1}y^3)^{-3} = 2^{-3}(x^{-1})^{-3}(y^3)^{-3} = \frac{1}{2^3} \cdot x^3 \cdot y^{-9} = \frac{1}{8} \cdot x^3 \cdot \frac{1}{y^9} = \frac{x^3}{8y^9}$.

65 $\frac{2^0 - 3^0}{4x^{-1}} = \frac{1-1}{4x^{-1}} = \frac{0 \cdot x}{4} = 0$.

69 $2^{-1} + 3^{-1} = \frac{1}{2} + \frac{1}{3} = \frac{3}{6} + \frac{2}{6} = \frac{5}{6}$.

73 $\frac{a - a^{-1}}{a^{-2} - a^{-3}} = \frac{a - \frac{1}{a}}{\frac{1}{a^2} - \frac{1}{a^3}} = \frac{\left(a - \frac{1}{a}\right) \cdot a^3}{\left(\frac{1}{a^2} - \frac{1}{a^3}\right) \cdot a^3} = \frac{a^4 - a^2}{a - 1} = \frac{a^2(a+1)(a-1)}{a-1} = a^2(a+1)$.

77 $x^{3n}x^{-5n} = x^{3n-5n} = x^{-2n} = \frac{1}{x^{2n}}$.

81 $\frac{1}{\left(\frac{5}{7}\right)^{-1}} = \frac{1}{\frac{7}{5}} = 1 \cdot \frac{5}{7} = \frac{5}{7}$.

85 $\left[\frac{x^{-n}x^{4n}}{(x^3)^n}\right]^{-1} = \frac{(x^3)^n}{x^{-n}x^{4n}} = \frac{x^{3n}}{x^{-n+4n}} = \frac{x^{3n}}{x^{3n}} = 1$.

89 $(2^{-1} - 3^{-1})^{-1} = \left(\frac{1}{2} - \frac{1}{3}\right)^{-1} = \left(\frac{3}{6} - \frac{2}{6}\right)^{-1} = \left(\frac{1}{6}\right)^{-1} = \frac{6}{1} = 6$.

93 $\frac{(2x^{-1}y^3)^3(3y^{-2}z^{-1})^2}{(x^2z^3)^{-1}} = \frac{2^3 x^{-3} y^9 \cdot 3^2 y^{-4} z^{-2}}{x^{-2} z^{-3}} = \frac{72 x^{-3} y^5 z^{-2}}{x^{-2} z^{-3}} = \frac{72 y^5 z}{x}$.

Review Problems

101 $\frac{2.051}{100} = .02051$

Exercises 8.2

1 To express 1.41×10^8 in decimal notation move the decimal point 8 places to the right. To do this we must add 6 zeros: $1.41 \times 10^8 = 141{,}000{,}000$.

5 To express 7.35×10^{25} as a decimal we must move the decimal point 25 places to the right which means we must add 23 zeros: $7.35 \times 10^{25} = 73{,}500{,}000{,}000{,}000{,}000{,}000{,}000{,}000$.

9 To express 3.5×10^{-12} as a decimal we must move the decimal point 12 places to the left which means that 11 zeros must be added to the left: $3.5 \times 10^{-12} = 0.000\,000\,000\,003\,5$.

13 To go from $36{,}000{,}000.$ to 3.6 requires that we move the decimal point 7 places to the left. Therefore $36{,}000{,}000 = 3.6 \times 10^7$.

17 To go from $970{,}000.$ to 9.7 requires the decimal point to be moved 5 places. Therefore $970{,}000 = 9.7 \times 10^5$.

21 To express $0.000\,001\,53$ as 1.53×10^{-n}, we find n by counting the number of places we must move the decimal point to go from $0.000\,001\,53$ to 1.53. Since we must move the decimal point 6 places, $0.000\,001\,53 = 1.53 \times 10^{-6}$.

25 $(3{,}000{,}000)(40{,}000{,}000) = (3 \times 10^6)(4 \times 10^7) = 3 \times 4 \times 10^{6+7} = 12 \times 10^{13} = 1.2 \times 10^1 \times 10^{13} = 1.2 \times 10^{14}$.

29 $\dfrac{133 \times 10^{-13}}{.001} = \dfrac{1.33 \times 10^2 \times 10^{-13}}{1 \times 10^{-3}} = \dfrac{1.33 \times 10^{-11}}{10^{-3}} = 1.33 \times 10^{-11} \times 10^3 = 1.33 \times 10^{\,8}$.

33 $\dfrac{(2000)^3(.001)}{.00004} = \dfrac{(2 \times 10^3)^3(1 \times 10^{-3})}{4 \times 10^{-5}} = \dfrac{8 \times 1}{4} \cdot \dfrac{10^9 \times 10^{-3}}{10^{-5}} = \dfrac{2 \times 10^6}{10^{-5}} = 2 \times 10^6 \times 10^5 = 2.0 \times 10^{11}$.

Exercises 8.2

37 $(30,000)^4 = (3 \times 10^4)^4 = 3^4 \times (10^4)^4 = 81 \times 10^{16} = 8.1 \times 10 \times 10^{16} = 8.1 \times 10^{17}$.

41 Since $T = \frac{D}{R}$ and the circumference of the earth is $24{,}800 = 2.48 \times 10^4$ and the speed of light is 186,000 mps $= 1.86 \times 10^5$, the number of seconds is
$$\frac{2.48 \times 10^4}{1.86 \times 10^5} = \frac{2.48}{1.86} \times 10^{-1} = 1.33 \times 10^{-1} = .133 \text{ sec }.$$

45 Since $D = RT$, the earth has traveled $(4.5 \times 10^9)(5.84 \times 10^8) = 26.3 \times 10^{17} = 2.63 \times 10^{18}$.

49 The mass of the sun is 1.99×10^{33} and the mass of the earth is 5.98×10^{27}. Therefore the sun is $\frac{1.99 \times 10^{33}}{5.98 \times 10^{27}} = .333 \times 10^6 = 3.33 \times 10^5$ more massive than the earth.

53 $(-6)^{17} \approx 1.69 \times 10^{13}$ **57** $(186,000)^2 = 3.46 \times 10^{10}$ **61** $\frac{18{,}000}{.003} = 6 \times 10^6$

65 75 years $= 75 \times 365$ days $= 75 \times 365 \times 24$ hours $= 75 \times 365 \times 24 \times 60$ minutes $= 75 \times 365 \times 24 \times 60 \times 60$ seconds $\approx 2.37 \times 10^9$ seconds.

69 $(10{,}000{,}000)^8(.000\ 001)^4 = (1 \times 10^7)^8(1 \times 10^{-6})^4 = 10^{56} \times 10^{-24} = 10^{56-24} = 1 \times 10^{32}$.

73 $(2{,}100{,}000)^3(4{,}000)^2 = (2.1 \times 10^6)^3(4 \times 10^3)^2 = (9.26 \times 10^{18})(16 \times 10^6) =$
$(9.26 \times 16) \times 10^{18+6} = 148.16 \times 10^{24} \approx 1.48 \times 10^2 \times 10^{24} = 1.48 \times 10^{26}$.

Review Problems

81 $|-6| = 6$ **85** The rational numbers are $8, -7, 0, \frac{1}{4}, .71$, and $.71\overline{71}$.

Exercises 8.3

1   16 has 2 square roots -4 and 4 because $(-4)^2 = 16$ and $4^2 = 16$.

5   Both 11 and -11 are square roots of 121 since $11^2 = 121$ and $(-11)^2 = 121$.

9 $\sqrt{0} = 0$ because $0^2 = 0$.   **13** $(\sqrt{25})^2 = 5^2 = 25$.

17 $\sqrt{6} \cdot \sqrt{6} = \sqrt{36} = 6$.   **21** $\sqrt{7^2} = \sqrt{49} = 7$.

25 $\sqrt{x^4} = x^2$ because $x^2 \cdot x^2 = x^4$. Therefore $-2\sqrt{x^4} = -2x^2$.

29 $-\sqrt{\dfrac{25}{a^2}} = -\dfrac{\sqrt{25}}{\sqrt{a^2}} = -\dfrac{5}{a}$.

33   Note that $(t^3)^2 = t^6$ has 2 square roots t^3 and $-t^3$ where $t^3 < 0$ and $-t^3 > 0$ since $t < 0$.
But $\sqrt{(t^3)^2}$ is the principal square root which means the non-negative square root.
Since $\sqrt{(t^3)^2}$ must be non-negative, $\sqrt{(t^3)^2} = -t^3$.

37   Since $\sqrt{A}$ is the principal square root, which means the non-negative square root, $\sqrt{x^2} = |x|$ which guarantees that the answer is non-negative.

41 $\sqrt{a^2 + 4a + 4} = \sqrt{(a+2)^2} = |a+2|$. Note we need the absolute value signs to make sure the square root is the non-negative one.

45 $\sqrt[3]{64} = 4$ because $4^3 = 64$. Therefore $2 \cdot \sqrt[3]{64} = 2 \cdot 4 = 8$.

49 $\sqrt[5]{-1} = -1$ since $(-1)^5 = -1$.   **53** $\sqrt[3]{\dfrac{8}{27}} = \dfrac{\sqrt[3]{8}}{\sqrt[3]{27}} = \dfrac{2}{3}$.

57 $10\sqrt{82} \approx 90.554$ **61** $\sqrt[3]{-25} \approx -2.924$

Exercises 8.3

[65] $8\sqrt{14} - 2\sqrt{14} = 6\sqrt{14} \approx 22.450$.

[69] $2 + \sqrt[4]{4} \approx 2 + 1.414 = 3.414$.

[73] 1.414214 is a rational number and a real number. It is not a natural number, integer, whole number, or irrational number.

[77] $\sqrt[3]{-9}$ is an irrational number and a real number. It is not a natural number, integer, or whole number.

[81] $-\sqrt{2} + \sqrt{2} = 0$. Therefore it is a whole number, an integer, a rational number, and a real number. It is not a natural number or an irrational number.

[85] $\sqrt{.0049} = \sqrt{\frac{49}{10,000}} = \frac{\sqrt{49}}{\sqrt{10,000}} = \frac{7}{100}$.

[89] $\sqrt{2\frac{1}{4}} = \sqrt{\frac{9}{4}} = \frac{\sqrt{9}}{\sqrt{4}} = \frac{3}{2}$.

[93] $\sqrt{49a^4} = \sqrt{(7a^2)^2} = 7a^2$.

Review Problems

[101] $114 = 2 \times 3 \times 19$

[105] $3y - 12 > 0 \Rightarrow 3y > 12 \Rightarrow y > 4$.

Exercises 8.4

[1] $\sqrt{28} = \sqrt{4 \times 7} = \sqrt{4}\sqrt{7} = 2\sqrt{7}$.

[5] $\sqrt{396b^5} = \sqrt{36 \cdot 11b^4 b} = \sqrt{36}\sqrt{b^4}\sqrt{11b} = 6b^2\sqrt{11b}$.

[9] $\sqrt{144 + 25} = \sqrt{169} = \sqrt{(13)^2} = 13$.

[13] $\sqrt{3}\sqrt{12} = \sqrt{3 \times 12} = \sqrt{36} = 6$.

[17] $ab\sqrt{a^2b}\sqrt{ab^2} = ab\sqrt{(a^2b)(ab^2)} = ab\sqrt{a^3b^3} = ab\sqrt{a^2b^2}\sqrt{ab} = ab(ab)\sqrt{ab} = a^2b^2\sqrt{ab}$.

Exercises 8.4

21 $\sqrt{8}\sqrt{6}\sqrt{21} = \sqrt{(4\times 2)(3\times 2)(7\times 3)} = \sqrt{16\times 9\times 7} = \sqrt{16}\sqrt{9}\sqrt{7} = 4\times 3\sqrt{7} = 12\sqrt{7}$.

25 $\sqrt[3]{10}\cdot\sqrt[3]{25c^3} = \sqrt[3]{(10)(25c^3)} = \sqrt[3]{250c^3} = \sqrt[3]{125\cdot 2c^3} = \sqrt[3]{125}\cdot\sqrt[3]{c^3}\cdot\sqrt[3]{2} = 5c\cdot\sqrt[3]{2}$.

29 $\sqrt[4]{75}\cdot\sqrt[4]{50} = \sqrt[4]{(5^2\cdot 3)(5^2\cdot 2)} = \sqrt[4]{5^4\cdot 6} = \sqrt[4]{5^4}\cdot\sqrt[4]{6} = 5\cdot\sqrt[4]{6}$.

33 $\sqrt[4]{27b^3}\cdot\sqrt[4]{3b^2} = \sqrt[4]{3^3\cdot 3b^5} = \sqrt[4]{3^4 b^4 b} = 3b\cdot\sqrt[4]{b}$.

37 $\sqrt{.000\,000\,36} = \sqrt{36\times 10^{-8}} = \sqrt{36}\sqrt{10^{-8}} = 6\times 10^{-4}$. Note that $\sqrt{10^{-8}} = 10^{-4}$ because $(10^{-4})^2 = 10^{-8}$.

41 $-\sqrt{\frac{2}{9}} = -\frac{\sqrt{2}}{\sqrt{9}} = -\frac{\sqrt{2}}{3}$.

45 $\frac{\sqrt{27}}{\sqrt{3}} = \sqrt{\frac{27}{3}} = \sqrt{9} = 3$.

49 $\sqrt[3]{\frac{2}{27}} = \frac{\sqrt[3]{2}}{\sqrt[3]{27}} = \frac{\sqrt[3]{2}}{3}$.

53 $\sqrt[5]{\frac{3}{a^5}} = \frac{\sqrt[5]{3}}{\sqrt[5]{a^5}} = \frac{\sqrt[5]{3}}{a}$.

57 $\sqrt{\frac{10xy}{18x^3}} = \sqrt{\frac{5y}{9x^2}} = \frac{\sqrt{5y}}{\sqrt{9x^2}} = \frac{\sqrt{5y}}{3x}$.

61 $\sqrt{\frac{a^8}{100}} = \frac{\sqrt{(a^4)^2}}{\sqrt{100}} = \frac{a^4}{10}$.

65 $\sqrt{x-7}$ is defined as a real number when $x-7\geq 0$ or $x\geq 7$. Therefore the domain is $\{x/x\geq 7\}$.

69 $\frac{z+25}{\sqrt{z^2+4}}$ is defined as a real number when $z^2+4 > 0$. But z^2+4 is always greater than 0. Therefore the domain is all real numbers.

73 $\sqrt{x^2+6x+9} = \sqrt{(x+3)^2} = |x+3|$. Note we need the absolute value signs to make sure that $x+3$ is non-negative.

77 $3\sqrt{x^3+6x^2} = 3\sqrt{x^2(x+6)} = 3x\sqrt{x+6}$.

81 $\frac{3}{\sqrt[5]{x^2+1}}$ is defined as long as $x^2+1\neq 0$. But x^2+1 is never 0. Therefore the domain is $\Re$.

Review Problems

[85] $7x + 3x = (7+3)x = 10x$.

[89] $(2x-y)(2x-y) = 4x^2 - 2xy - 2xy + y^2 = 4x^2 - 4xy + y^2$.

[93] $\dfrac{(x-9)(y+3)}{x-9} = y+3$

Exercises 8.5

[1] $\dfrac{1}{\sqrt{5}} = \dfrac{1}{\sqrt{5}} \cdot \dfrac{\sqrt{5}}{\sqrt{5}} = \dfrac{\sqrt{5}}{5}$.

[5] $\dfrac{\sqrt{3}}{\sqrt{6}} = \dfrac{\sqrt{3}}{\sqrt{6}} \cdot \dfrac{\sqrt{6}}{\sqrt{6}} = \dfrac{\sqrt{18}}{6} = \dfrac{\sqrt{9 \times 2}}{6} = \dfrac{3\sqrt{2}}{6} = \dfrac{\sqrt{2}}{2}$.

[9] $\sqrt[3]{\dfrac{1}{3}} = \dfrac{\sqrt[3]{1}}{\sqrt[3]{3}} = \dfrac{1}{\sqrt[3]{3}} \cdot \dfrac{\sqrt[3]{9}}{\sqrt[3]{9}} = \dfrac{\sqrt[3]{9}}{\sqrt[3]{27}} = \dfrac{\sqrt[3]{9}}{3}$.

[13] $5\sqrt{7} + 4\sqrt{7} = (5+4)\sqrt{7} = 9\sqrt{7}$.

[17] $\sqrt{12} - \sqrt{75} + \sqrt{5} = \sqrt{4 \times 3} - \sqrt{25 \times 3} + \sqrt{5} = 2\sqrt{3} - 5\sqrt{3} + \sqrt{5} = (2-5)\sqrt{3} + \sqrt{5} = -3\sqrt{3} + \sqrt{5}$.

[21] $\sqrt{\dfrac{3}{4}}\sqrt{\dfrac{8}{15}} - \sqrt{40} = \sqrt{\dfrac{3}{4} \cdot \dfrac{8}{15}} - \sqrt{4 \cdot 10} = \sqrt{\dfrac{2}{5}} - 2\sqrt{10} = \dfrac{\sqrt{2}}{\sqrt{5}} \cdot \dfrac{\sqrt{5}}{\sqrt{5}} - 2\sqrt{10} = \dfrac{\sqrt{10}}{5} - 2\sqrt{10} = \dfrac{\sqrt{10}}{5} - \dfrac{10\sqrt{10}}{5} = \dfrac{\sqrt{10} - 10\sqrt{10}}{5} = \dfrac{(1-10)\sqrt{10}}{5} = -\dfrac{9\sqrt{10}}{5}$.

[25] $(2\sqrt{5})(3\sqrt{7}) = (2 \cdot 3)(\sqrt{5} \cdot \sqrt{7}) = 6\sqrt{35}$.

[29] $(2+\sqrt{3})(3+\sqrt{5}) = 6 + 2\sqrt{5} + 3\sqrt{3} + \sqrt{3}\sqrt{5} = 6 + 2\sqrt{5} + 3\sqrt{3} + \sqrt{15}$.

[33] $(2+\sqrt{7})(2-\sqrt{7}) = 4 - 2\sqrt{7} + 2\sqrt{7} - \sqrt{7}\sqrt{7} = 4 - 7 = -3$.

Exercises 8.5

37 The conjugate of $4 - \sqrt{3}$ is $4 + \sqrt{3}$.

41 $\dfrac{10}{1+\sqrt{6}} = \dfrac{10}{1+\sqrt{6}} \cdot \dfrac{1-\sqrt{6}}{1-\sqrt{6}} = \dfrac{10(1-\sqrt{6})}{1-6} = \dfrac{10(1-\sqrt{6})}{-5} = -2(1-\sqrt{6}) = -2 + 2\sqrt{6}$.

45 $\dfrac{1+\sqrt{3}}{1-\sqrt{3}} = \dfrac{1+\sqrt{3}}{1-\sqrt{3}} \cdot \dfrac{1+\sqrt{3}}{1+\sqrt{3}} = \dfrac{1+\sqrt{3}+\sqrt{3}+3}{1-3} = \dfrac{4+2\sqrt{3}}{-2} = \dfrac{2(2+\sqrt{3})}{-2} = -1(2+\sqrt{3}) =$
$-2-\sqrt{3}$.

49 $\dfrac{\sqrt{15}}{\sqrt{3}} = \sqrt{\dfrac{15}{3}} = \sqrt{5}$.   **53** $\dfrac{3}{2-\sqrt{5}} = \dfrac{3}{2-\sqrt{5}} \cdot \dfrac{2+\sqrt{5}}{2+\sqrt{5}} = \dfrac{3(2+\sqrt{5})}{4-5} = \dfrac{6+3\sqrt{5}}{-1} =$
$-6-3\sqrt{5}$.

57 $\dfrac{30x^9}{\sqrt{12x}} = \dfrac{30x^9}{\sqrt{4\cdot 3x}} = \dfrac{30x^9}{2\sqrt{3x}} \cdot \dfrac{\sqrt{3x}}{\sqrt{3x}} = \dfrac{30x^9 \cdot \sqrt{3x}}{2\cdot 3x} = 5x^8 \cdot \sqrt{3x}$.

61 $3\sqrt{72} - 2\sqrt{242} = 3\sqrt{36\times 2} - 2\sqrt{121\times 2} = 3\times 6\sqrt{2} - 2\times 11\sqrt{2} = 18\sqrt{2} - 22\sqrt{2} = -4\sqrt{2}$.

65 $\dfrac{\sqrt{2}+\sqrt{3}}{\sqrt{5}+2\sqrt{2}} = \dfrac{\sqrt{2}+\sqrt{3}}{\sqrt{5}+2\sqrt{2}} \cdot \dfrac{\sqrt{5}-2\sqrt{2}}{\sqrt{5}-2\sqrt{2}} = \dfrac{(\sqrt{2}+\sqrt{3})(\sqrt{5}-2\sqrt{2})}{5-4\cdot 2} =$
$\dfrac{\sqrt{10}-2\sqrt{2}\sqrt{2}+\sqrt{15}-2\sqrt{3}\sqrt{2}}{5-8} = \dfrac{\sqrt{10}-4+\sqrt{15}-2\sqrt{6}}{-3} \cdot \dfrac{-1}{-1} = \dfrac{-\sqrt{10}+4-\sqrt{15}+2\sqrt{6}}{3}$.

69 $\dfrac{\sqrt[3]{16}+\sqrt[3]{54}}{\sqrt[3]{2}} = \dfrac{\sqrt[3]{16}+\sqrt[3]{54}}{\sqrt[3]{2}} \cdot \dfrac{\sqrt[3]{2^2}}{\sqrt[3]{2^2}} = \dfrac{\sqrt[3]{16\cdot 4}+\sqrt[3]{54\cdot 4}}{\sqrt[3]{2^3}} = \dfrac{\sqrt[3]{64}+\sqrt[3]{216}}{\sqrt[3]{8}} = \dfrac{4+6}{2} = 5$.

73 $\left(\dfrac{1+\sqrt{3}}{2}\right)^2 = \dfrac{(1+\sqrt{3})(1+\sqrt{3})}{2^2} = \dfrac{1+\sqrt{3}+\sqrt{3}+3}{4} = \dfrac{4+2\sqrt{3}}{4} = \dfrac{2(2+\sqrt{3})}{4} = \dfrac{2+\sqrt{3}}{2}$.

77 $2\sqrt{125} - \sqrt{50} + \sqrt{245} = 2\sqrt{25\times 5} - \sqrt{25\times 2} + \sqrt{49\times 5} = 2\cdot 5\sqrt{5} - 5\sqrt{2} + 7\sqrt{5} =$
$10\sqrt{5} - 5\sqrt{2} + 7\sqrt{5} = 17\sqrt{5} - 5\sqrt{2}$.

Exercises 8.5

81 $(\sqrt{5}+2\sqrt{3})(\sqrt{10}-\sqrt{3}) = \sqrt{50}-\sqrt{15}+2\sqrt{30}-2\sqrt{9} = \sqrt{25\cdot 2}-\sqrt{15}+2\sqrt{30}-6 =$
$5\sqrt{2}-\sqrt{15}+2\sqrt{30}-6$.

85 $2\sqrt{\frac{1}{8}}-3\sqrt{\frac{1}{2}} = 2\cdot\frac{\sqrt{1}}{\sqrt{8}}-3\cdot\frac{\sqrt{1}}{\sqrt{2}} = \frac{2}{\sqrt{8}}\cdot\frac{\sqrt{2}}{\sqrt{2}}-\frac{3}{\sqrt{2}}\cdot\frac{\sqrt{2}}{\sqrt{2}} = \frac{2\sqrt{2}}{\sqrt{16}}-\frac{3\sqrt{2}}{\sqrt{4}} = \frac{2\sqrt{2}}{4}-\frac{3\sqrt{2}}{2} =$
$\frac{\sqrt{2}}{2}-\frac{3\sqrt{2}}{2} = \frac{\sqrt{2}-3\sqrt{2}}{2} = \frac{-2\sqrt{2}}{2} = -\sqrt{2}$.

89 $\frac{12}{\sqrt[4]{18}} = \frac{12}{\sqrt[4]{3^2\cdot 2}} = \frac{12}{\sqrt[4]{3^2\cdot 2}}\cdot\frac{\sqrt[4]{3^2\cdot 2^3}}{\sqrt[4]{3^2\cdot 2^3}} = \frac{12\cdot\sqrt[4]{72}}{\sqrt[4]{3^4\cdot 2^4}} = \frac{12\cdot\sqrt[4]{72}}{3\cdot 2} = 2\cdot\sqrt[4]{72}$.

93 $(\sqrt{2}+1)^3 = (\sqrt{2}+1)^2(\sqrt{2}+1) = (2+2\sqrt{2}+1)(\sqrt{2}+1) = (2\sqrt{2}+3)(\sqrt{2}+1) =$
$2\cdot 2+2\sqrt{2}+3\sqrt{2}+3 = 7+5\sqrt{2}$.

97 $\frac{1}{3+\sqrt{3}}+\frac{1}{\sqrt{3}-3} = \frac{1}{3+\sqrt{3}}\cdot\frac{3-\sqrt{3}}{3-\sqrt{3}}+\frac{1}{\sqrt{3}-3}\cdot\frac{\sqrt{3}+3}{\sqrt{3}+3} = \frac{3-\sqrt{3}}{9-3}+\frac{\sqrt{3}+3}{3-9} =$
$\frac{3-\sqrt{3}}{6}+\frac{\sqrt{3}+3}{-6} = \frac{3-\sqrt{3}}{6}+\frac{\sqrt{3}+3}{-6}\cdot\frac{-1}{-1} = \frac{3-\sqrt{3}}{6}+\frac{-\sqrt{3}-3}{6} = \frac{3-\sqrt{3}-\sqrt{3}-3}{6} =$
$\frac{-2\sqrt{3}}{6} = \frac{-\sqrt{3}}{3}$.

Review Problems

103 $3x+4=-8 \Rightarrow 3x=-12 \Rightarrow x=-4$.

107 $A=P+Prt \Rightarrow A=P(1+rt) \Rightarrow P=\frac{A}{1+rt}$.

111 $\frac{z}{x+y}=\frac{2}{3} \Rightarrow 3z=2x+2y \Rightarrow 3z-2x=2y \Rightarrow y=\frac{3z-2x}{2}$.

Exercises 8.6

1. To solve $\sqrt{x} = 3$ square both sides. Then $(\sqrt{x})^2 = 3^2 \Rightarrow x = 9$. We must check this answer to see if it is an extraneous solution.

   Check: $\quad \sqrt{9} = 3$

   $\qquad\qquad 3 = 3 \quad \checkmark$

   Therefore $x = 9$ is the solution.

5. To solve $-\sqrt{3t-2} = 5$ square both sides. Then $(-\sqrt{3t-2})^2 = 5^2 \Rightarrow$
   $3t - 2 = 25 \Rightarrow 3t = 27 \Rightarrow t = 9$. But we must must check this solution in the original equation: $\quad -\sqrt{3 \cdot 9 - 2} = 5$

   $\qquad\qquad -\sqrt{27 - 2} = 5$

   $\qquad\qquad -\sqrt{25} = 5$

   $\qquad\qquad -5 \neq 5$

   Therefore $t = 9$ is an extraneous solution and the equation has no solutions.

9. $\sqrt{3x-4} = \sqrt{11-2x} \Rightarrow (\sqrt{3x-4})^2 = (\sqrt{11-2x})^2 \Rightarrow 3x - 4 = 11 - 2x \Rightarrow 5x = 15 \Rightarrow$
   $x = 3$. Check: $\quad \sqrt{3 \cdot 3 - 4} = \sqrt{11 - 2 \cdot 3}$

   $\qquad\qquad \sqrt{5} = \sqrt{5} \quad \checkmark$

   Therefore $x = 3$ is the solution.

13. $1 + \sqrt{2x+1} = 6 \Rightarrow \sqrt{2x+1} = 5 \Rightarrow (\sqrt{2x+1})^2 = 5^2 \Rightarrow 2x + 1 = 25 \Rightarrow$
    $2x = 24 \Rightarrow x = 12$.

    Check $x = 12$: $\quad 1 + \sqrt{(2 \cdot 12) + 1} = 6$

    $\qquad\qquad 1 + \sqrt{2 \cdot 12 + 1} = 6$

    $\qquad\qquad 1 + \sqrt{25} = 6$

    $\qquad\qquad 1 + 5 = 6$

    $\qquad\qquad 6 = 6 \quad \checkmark$

    Therefore $x = 12$ is a solution.

Exercises 8.6

17 $1-x = \sqrt{x-1} \Rightarrow (1-x)^2 = (\sqrt{x-1})^2 \Rightarrow 1-2x+x^2 = x-1 \Rightarrow x^2-3x+2=0 \Rightarrow$
$(x-2)(x-1)=0 \Rightarrow x = 1 \text{ or } 2$.

Check $x=1$: $1-1 = \sqrt{1-1}$ Check $x=2$: $1-2 = \sqrt{2-1}$
                      $0 = 0$ ✓ $-1 = \sqrt{1}$
                                                                                   $-1 = 1$ ✗

Therefore $x = 1$ is the only solution.

21 $\sqrt{3t} - 2 = t - 8 \Rightarrow \sqrt{3t} = t - 6 \Rightarrow (\sqrt{3t})^2 = (t-6)^2 \Rightarrow 3t = t^2 - 12t + 36 \Rightarrow$
$0 = t^2 - 15t + 36 \Rightarrow (t-12)(t-3) = 0 \Rightarrow t = 12 \text{ or } 3$.

Check $t=12$: $\sqrt{3 \cdot 12} - 2 = 12 - 8$ Check $t=3$: $\sqrt{3 \cdot 3} - 2 = 3 - 8$
                        $\sqrt{36} - 2 = 4$ $\sqrt{9} - 2 = -5$
                        $6 - 2 = 4$ $3 - 2 = -5$
                        $4 = 4$ ✓ $1 = -5$ ✗

Therefore $t = 12$ is the only solution.

25 $\sqrt{\frac{5a}{4}} - 3 = 7 \Rightarrow \sqrt{\frac{5a}{4}} = 10 \Rightarrow \left(\sqrt{\frac{5a}{4}}\right)^2 = 10^2 \Rightarrow \frac{5a}{4} = 100 \Rightarrow 5a = 400 \Rightarrow x = 80$.

Check: $\sqrt{\frac{5 \cdot 80}{4}} - 3 = 7$
            $\sqrt{5 \cdot 20} - 3 = 7$
            $\sqrt{100} - 3 = 7$
            $10 - 3 = 7$
            $7 = 7$ ✓   Therefore $x = 7$ is the solution.

29 $\sqrt{x^3 + 11x} = \sqrt{5x}\sqrt{x+1} \Rightarrow \sqrt{x^3 + 11x} = \sqrt{5x(x+1)} \Rightarrow$
$(\sqrt{x^3 + 11x})^2 = (\sqrt{5x(x+1)})^2 \Rightarrow x^3 + 11x = 5x^2 + 5x \Rightarrow x^3 - 5x^2 + 6x = 0 \Rightarrow$
$x(x^2 - 5x + 6) = 0 \Rightarrow x(x-3)(x-2) = 0 \Rightarrow x = 0, 2, \text{ or } 3$.

Check $x = 0$: $\sqrt{0+0} = \sqrt{0} \cdot \sqrt{1}$ Check $x = 2$: $\sqrt{8 + 22} = \sqrt{10} \cdot \sqrt{3}$
                          $0 = 0$ ✓ $\sqrt{30} = \sqrt{30}$ ✓

Check $x = 3$: $\sqrt{27 + 33} = \sqrt{15} \cdot \sqrt{4}$
                     $\sqrt{60} = \sqrt{60}$ ✓   Therefore $x = 0, 2, \text{ and } 3$ are solutions.

Exercises 8.6

33 $\sqrt[5]{11x-1} = 2 \Rightarrow \left(\sqrt[5]{11x-1}\right)^5 = 2^5 \Rightarrow 11x - 1 = 32 \Rightarrow 11x = 33 \Rightarrow x = 3$.
The solution is $x = 3$.

37 $4 + \sqrt[3]{m-1} = 0 \Rightarrow \sqrt[3]{m-1} = -4 \Rightarrow \left(\sqrt[3]{m-1}\right)^3 = (-4)^3 \Rightarrow m - 1 = -64 \Rightarrow m = -63$. Therefore the solution is $m = -63$.

41 $r = \sqrt{\frac{A}{\pi}} \Rightarrow r^2 = \left(\sqrt{\frac{A}{\pi}}\right)^2 \Rightarrow r^2 = \frac{A}{\pi} \Rightarrow A = \pi r^2$.

45 $x = \sqrt{\frac{kM}{2f}} \Rightarrow x^2 = \left(\sqrt{\frac{kM}{2f}}\right)^2 \Rightarrow x^2 = \frac{kM}{2f} \Rightarrow 2fx^2 = kM \Rightarrow f = \frac{kM}{2x^2}$.

49 $r = \sqrt{\frac{2A}{x}} \Rightarrow r^2 = \left(\sqrt{\frac{2A}{x}}\right)^2 \Rightarrow r^2 = \frac{2A}{x} \Rightarrow xr^2 = 2A \Rightarrow A = \frac{xr^2}{2}$.

53 $r = \sqrt{\frac{3V}{\pi h}} \Rightarrow r^2 = \frac{3V}{\pi h} \Rightarrow \pi h r^2 = 3V \Rightarrow h = \frac{3V}{\pi r^2}$.

57 $y = Bx^2 \sqrt{\frac{q}{8mv}} \Rightarrow y^2 = \left(Bx^2 \sqrt{\frac{q}{8mv}}\right)^2 \Rightarrow y^2 = B^2 x^4 \cdot \frac{q}{8mv} \Rightarrow 8mvy^2 = B^2 x^4 q \Rightarrow q = \frac{8mvy^2}{B^2 x^4}$.

61 $k = 3 + 2\sqrt{x+1} \Rightarrow k - 3 = 2\sqrt{x+1} \Rightarrow (k-3)^2 = (2\sqrt{x+1})^2 \Rightarrow k^2 - 6k + 9 = 4(x+1) \Rightarrow k^2 - 6k + 9 = 4x + 4 \Rightarrow k^2 - 6k + 5 = 4x \Rightarrow x = \frac{k^2 - 6k + 5}{4}$.

65 $3x \cdot \sqrt[3]{2xy} = 9x^5 \Rightarrow \sqrt[3]{2xy} = \frac{9x^5}{3x} \Rightarrow \sqrt[3]{2xy} = 3x^4 \Rightarrow \left(\sqrt[3]{2xy}\right)^3 = (3x^4)^3 \Rightarrow 2xy = 27x^{12} \Rightarrow y = \frac{27x^{12}}{2x} \Rightarrow y = \frac{27x^{11}}{2}$.

69 $\sqrt[5]{a+b+c} = -1 \Rightarrow \left(\sqrt[5]{a+b+c}\right)^5 = (-1)^5 \Rightarrow a + b + c = -1 \Rightarrow b = -a - c - 1 \text{ or } b = -(a + c + 1)$.

Exercises 8.6

73 Problem Statement: The square root of 6 more than 3 times a number is the same as 3 times the square root of 3. Find the number.

Let x be the number. Then $\sqrt{3x+6} = 3\sqrt{3} \Rightarrow (\sqrt{3x+6})^2 = (3\sqrt{3})^2 \Rightarrow$
$3x + 6 = 9 \cdot 3 \Rightarrow 3x + 6 = 27 \Rightarrow 3x = 21 \Rightarrow x = 7$.

Check: $\sqrt{3 \cdot 7 + 6} = 3\sqrt{3}$
$\sqrt{27} = 3\sqrt{3}$
$\sqrt{9 \cdot 3} = 3\sqrt{3}$
$3\sqrt{3} = 3\sqrt{3}$ ✓ Therefore the number is 7.

77 If $T = 2\pi\sqrt{\frac{L}{32}}$ and $L = 1$ foot then $T = 2\pi\sqrt{\frac{1}{32}} = 2(3.1416)\sqrt{\frac{1}{32}} = 1.11$.

81 If $T = 2$ seconds then $2 = 2\pi\sqrt{\frac{L}{32}} \Rightarrow \frac{2}{2\pi} = \sqrt{\frac{L}{32}} \Rightarrow \left(\frac{1}{\pi}\right)^2 = \left(\sqrt{\frac{L}{32}}\right)^2 \Rightarrow \frac{1}{\pi^2} = \frac{L}{32} \Rightarrow$
$L = \frac{32}{\pi^2} = 3.24$.

85 If $r = \frac{\sqrt{A} - \sqrt{P}}{\sqrt{P}}$ and $r = 7\%$ and $A = \$572$ then $.07 = \frac{\sqrt{572} - \sqrt{P}}{\sqrt{P}} \Rightarrow$
$.07\sqrt{P} = \sqrt{572} - \sqrt{P} \Rightarrow .07\sqrt{P} + \sqrt{P} = \sqrt{572} \Rightarrow 1.07\sqrt{P} = \sqrt{572} \Rightarrow$
$\sqrt{P} = \frac{\sqrt{572}}{1.07} \Rightarrow P = \left(\frac{\sqrt{572}}{1.07}\right)^2 = \frac{572}{(1.07)^2} \approx \$500$.

89 If $d = \frac{1}{2}\sqrt{6h}$ and $h = 54$ feet then $d = \frac{1}{2}\sqrt{6 \cdot 54} = \frac{1}{2}\sqrt{324} = 9$ miles.

93 If $r = \sqrt[3]{\frac{3V}{4\pi}}$ and $V = \frac{9\pi}{2}$ in^3 then $r = \sqrt[3]{\frac{3 \cdot (9\pi/2)}{4\pi}} = \sqrt[3]{\frac{3 \cdot (9\pi/2) \cdot 2}{4\pi \cdot 2}} = \sqrt[3]{\frac{27\pi}{8\pi}} =$
$\sqrt[3]{\frac{27}{8}} = \frac{3}{2}$ inches.

97 $\sqrt{x+6} = 2 + \sqrt{x} \Rightarrow (\sqrt{x+6})^2 = (2 + \sqrt{x})^2 \Rightarrow x + 6 = 4 + 4\sqrt{x} + x \Rightarrow$
$2 = 4\sqrt{x} \Rightarrow 1 = 2\sqrt{x} \Rightarrow 1^2 = (2\sqrt{x})^2 \Rightarrow 1 = 4x \Rightarrow x = \frac{1}{4}$.

Check: $\sqrt{\frac{1}{4} + 6} = 2 + \sqrt{\frac{1}{4}}$
$\sqrt{\frac{25}{4}} = 2 + \frac{1}{2}$ or $\frac{5}{2} = \frac{5}{2}$ ✓ Therefore the solution is $x = \frac{1}{4}$.

Exercises 8.6

101 $\sqrt{3x} + \sqrt{75x} = 18 \Rightarrow \sqrt{3x} + \sqrt{25 \cdot 3x} = 18 \Rightarrow \sqrt{3x} + 5\sqrt{3x} = 18 \Rightarrow 6\sqrt{3x} = 18 \Rightarrow$
$\sqrt{3x} = 3 \Rightarrow 3x = 9 \Rightarrow x = 3$.
Check: $\quad \sqrt{3 \cdot 3} + \sqrt{75 \cdot 3} = 18$
$\sqrt{9} + \sqrt{225} = 18$
$3 + 15 = 18$ or $18 = 18$ $\checkmark$. Therefore $x = 3$ is the solution.

105 $x\sqrt{3} + \sqrt{6} = \sqrt{150} - x\sqrt{27} \Rightarrow x\sqrt{3} + \sqrt{6} = \sqrt{25 \cdot 6} - x\sqrt{9 \cdot 3} \Rightarrow$
$x\sqrt{3} + \sqrt{6} = 5\sqrt{6} - 3x\sqrt{3} \Rightarrow 4x\sqrt{3} = 4\sqrt{6} \Rightarrow x\sqrt{3} = \sqrt{6} \Rightarrow (x\sqrt{3})^2 = (\sqrt{6})^2 \Rightarrow$
$x^2 \cdot 3 = 6 \Rightarrow x^2 = 2 \Rightarrow x = \pm\sqrt{2}$.
Check $x = \sqrt{2}$: $\quad \sqrt{2}\sqrt{3} + \sqrt{6} = \sqrt{150} - \sqrt{2}\sqrt{27}$
$\sqrt{6} + \sqrt{6} = \sqrt{25 \cdot 6} - \sqrt{54}$
$2\sqrt{6} = 5\sqrt{6} - 3\sqrt{6}$
$2\sqrt{6} = 2\sqrt{6}$ $\checkmark$
Check $x = -\sqrt{2}$: $\quad -\sqrt{2}\sqrt{3} + \sqrt{6} = \sqrt{150} - (-\sqrt{2})\sqrt{27}$
$-\sqrt{6} + \sqrt{6} = \sqrt{150} + \sqrt{54}$
$0 = \sqrt{150} + \sqrt{54}$ X. Therefore $x = -\sqrt{2}$ is not a solution.
The only solution is $x = \sqrt{2}$.

Review Problems

109 $a^2 = 16 \Rightarrow a = \pm\sqrt{16} \Rightarrow a = \pm 4$.

113 If $t = \frac{1}{4}\sqrt{d}$ and $d = 16$ then $t = \frac{1}{4}\sqrt{16} = \frac{1}{4} \cdot 4 = 1$.

Exercises 8.7

1 If $a = 6$ and $b = 8$ then $c^2 = a^2 + b^2 = 6^2 + 8^2 = 36 + 64 = 100$. Therefore $c = \sqrt{100} = 10$.

5 If $a = 6$ and $c = \sqrt{72}$ then $c^2 = a^2 + b^2 \Rightarrow (\sqrt{72})^2 = 6^2 + b^2 \Rightarrow 72 = 36 + b^2 \Rightarrow$
$72 - 36 = b^2 \Rightarrow b^2 = 36 \Rightarrow b^2 = \sqrt{36} = 6$.

Exercises 8.7

[9] If $a = 2\sqrt{3}$ and $b = 3\sqrt{2}$ then $c^2 = (2\sqrt{3})^2 + (3\sqrt{2})^2 = 4 \times 3 + 9 \times 2 = 12 + 18 = 30 \Rightarrow c = \sqrt{30}$.

[13] $(4\sqrt{2})^2 = (2x)^2 + (2x)^2 \Rightarrow 16 \cdot 2 = 4x^2 + 4x^2 \Rightarrow 32 = 8x^2 \Rightarrow 4 = x^2 \Rightarrow x = 2$.

[17] $(x+1)^2 + (\sqrt{13})^2 = (x+2)^2 \Rightarrow x^2 + 2x + 1 + 13 = x^2 + 4x + 4 \Rightarrow 10 = 2x \Rightarrow x = 5$.

[21] The 3 numbers (6, 8, 9) are the lengths of the sides of a right triangle if $c^2 = a^2 + b^2$ where c is the largest number. Therefore we must check if $9^2 = 6^2 + 8^2$. $9^2 = 81$ and $6^2 + 8^2 = 36 + 64 = 100$. Since $81 \neq 100$ the 3 numbers are not the lengths of the sides of a right triangle.

[25] $20^2 = 400$, $21^2 = 441$, and $30^2 = 900$. Since $400 + 441 \neq 900$, the 3 numbers (20, 21, 30) are not the lengths of the sides of a right triangle.

[29] $65^2 + 72^2 = 4225 + 5184 = 9409$ and $97^2 = 9409$. Therefore the numbers (65, 72, 97) are the lengths of the sides of a right triangle.

[33] There are 16 small triangles. Note that the number of triangles in the square with side a plus the number of triangles in the square with side b is equal to the number of triangles in the square with side c. This illustrates that $a^2 + b^2 = c^2$.

[37] Since $a = 4$ then $c = 2a = 8$. Then $8^2 = 4^2 + b^2 \Rightarrow 64 = 16 + b^2 \Rightarrow 48 = b^2 \Rightarrow b = \sqrt{48} = \sqrt{16 \times 3} = 4\sqrt{3}$.

[41] Problem Statement: A leg and the hypotenuse of a right triangle are consecutive odd integers. The other leg is 8 cm. Find the other two sides.

Let x be the length of the leg and $x + 2$ the length of the hypotenuse. Then
$8^2 + x^2 = (x+2)^2 \Rightarrow 64 + x^2 = x^2 + 4x + 4 \Rightarrow 60 = 4x \Rightarrow x = 15$.
Therefore the leg is 15 cm and the hypotenuse is 17 cm.

[45] Problem Statement: A vertical pole is 20 feet high. It is to be supported by a wire running from the top of the pole to a point on the ground 21 feet from the base of the pole. How long a piece of wire is needed?

Let x be the length of the wire. Then x is the hypotenuse of a right triangle with base 21 feet and height 20 feet. Therefore
$$x^2 = 20^2 + 21^2 \Rightarrow x^2 = 400 + 441 \Rightarrow x^2 = 841 \Rightarrow x = \sqrt{841} = 29 \text{ feet}.$$

[49] Problem Statement: A doorway is 6.5 feet high and 2.5 feet wide. Can a square sheet of tin 7 feet on a side be carried through the doorway? Justify your answer.

The piece of tin can be carried through the doorway if the diagonal of the doorway is more than 7 feet. Let x be the diagonal of the doorway. Then
$$x = \sqrt{(6.5)^2 + (2.5)^2} = 6.96 \text{ feet}. \quad \text{Since } 7 > 6.96, \text{ the doorway is too small}.$$

[53] If y varies directly as the cube of x then $y = kx^3$. Since $x = -3$ when $y = 27$,
$27 = k(-3)^3 \Rightarrow 27 = -27k \Rightarrow k = -1$. Therefore $y = -x^3$.
To find x when $y = -\frac{3}{2}$, replace y by $-\frac{3}{2}$ and solve for x.
$$-\frac{3}{2} = -x^3 \Rightarrow \frac{3}{2} = x^3 \Rightarrow x = \sqrt[3]{\frac{3}{2}} = \sqrt[3]{\frac{3}{2} \cdot \frac{4}{4}} = \frac{\sqrt[3]{12}}{\sqrt[3]{8}} = \frac{\sqrt[3]{12}}{2}.$$

[57] If s varies directly as the cube root of V then $s = k \cdot \sqrt[3]{V}$. Since $s = 5$ when $V = 125$,
$5 = k \cdot \sqrt[3]{125} \Rightarrow 5 = 5k \Rightarrow k = 1$. Therefore $s = \sqrt[3]{V}$. When $s = 2 \cdot \sqrt[3]{2}$ then
$2 \cdot \sqrt[3]{2} = \sqrt[3]{V} \Rightarrow \left(2 \cdot \sqrt[3]{2}\right)^3 = \left(\sqrt[3]{V}\right)^3 \Rightarrow 8 \times 2 = V \Rightarrow V = 16$.

[61] If y varies inversely as x then $xy = k$. If $x = 2$ when $y = \frac{1}{8}$ then $2 \times \frac{1}{8} = k \Rightarrow k = \frac{1}{4}$.
Therefore $xy = \frac{1}{4}$. When $y = 8$ then $x \cdot 8 = \frac{1}{4} \Rightarrow 8x = \frac{1}{4} \Rightarrow x = \frac{1}{4} \cdot \frac{1}{8} = \frac{1}{32}$.

Exercises 8.7

65 If y varies inversely as the square root of x then $y \cdot \sqrt{x} = k$. If $y = -\frac{\sqrt{2}}{2}$ when $x = 18$ then
$-\frac{\sqrt{2}}{2} \cdot \sqrt{18} = k \Rightarrow k = -\frac{\sqrt{36}}{2} = -\frac{6}{2} = -3$. Therefore $y \cdot \sqrt{x} = -3$.
If $x = 12$ then $y \cdot \sqrt{12} = -3 \Rightarrow y = -\frac{3}{\sqrt{12}} = -\frac{3}{\sqrt{12}} \cdot \frac{\sqrt{3}}{\sqrt{3}} = -\frac{3\sqrt{3}}{\sqrt{36}} = -\frac{3\sqrt{3}}{6} = -\frac{\sqrt{3}}{2}$.

69 $IR = k$ where I is the current and R is the resistance. If $I = 50$ amps when $R = 12$ ohms then
$50 \times 12 = k \Rightarrow k = 600$. Then $IR = 600$. Therefore when the current is 60 amps
$60 \cdot R = 600 \Rightarrow R = \frac{600}{60} = 10$ ohms.

73 If the surface area of a square varies directly with the square of its radius then $S = kr^2$.
Since the surface area is 36π when $r = 3$, $36\pi = k \cdot 3^2 \Rightarrow 36\pi = 9k \Rightarrow k = \frac{36\pi}{9} = 4\pi$.
Now we know $S = 4\pi r^2$ and when $r = 6$, the surface area is $4\pi(6)^2 = 4\pi(36) = 144\pi$.

77 Problem Statement: Larry is standing on a dock, and he is pulling in a boat at a point 9 feet above the water with a 41 foot rope. When he has pulled in 26 feet of rope, how far has he moved the boat?

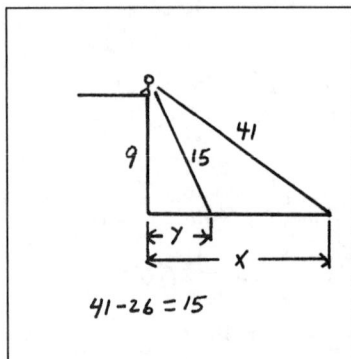

Let x be the initial distance of the boat from the dock, and let y be the distance from the dock after pulling in 26 feet of rope. See diagram at left. Then
$x^2 + 9^2 = 41^2 \Rightarrow x = \sqrt{41^2 - 9^2} = 40$ ft and
$y^2 + 9^2 = 15^2 \Rightarrow y = \sqrt{15^2 - 9^2} = 12$ ft.
Therefore the boat has moved $40 - 12 = 28$ feet.

Chapter 8 Review Exercises and Test

Chapter Review Exercises

1. $(x^{-3}y^4)^{-2} = x^6 y^{-8} = x^6 \cdot \dfrac{1}{y^8} = \dfrac{x^6}{y^8}$.

5. $\dfrac{x^{-1}+y^{-1}}{x+y} = \dfrac{\frac{1}{x}+\frac{1}{y}}{x+y} = \dfrac{\left(\frac{1}{x}+\frac{1}{y}\right)xy}{(x+y)xy} =$
$\dfrac{y+x}{(x+y)xy} = \dfrac{1}{xy}$.

9. $\sqrt[5]{-32} = \sqrt[5]{(-2)^5} = -2$.

13. $\dfrac{\sqrt{2x^{12}}}{\sqrt{18x^4}} = \sqrt{\dfrac{2x^{12}}{18x^4}} = \sqrt{\dfrac{x^8}{9}} = \sqrt{\dfrac{(x^4)^2}{9}} = \dfrac{\sqrt{(x^4)^2}}{\sqrt{9}} = \dfrac{x^4}{3}$.

17. $6\sqrt{\dfrac{1}{2}} + 3\sqrt{50} = 6 \cdot \dfrac{1}{\sqrt{2}} \cdot \dfrac{\sqrt{2}}{\sqrt{2}} + 3\sqrt{25 \times 2} = \dfrac{6\sqrt{2}}{2} + 3 \times 5\sqrt{2} = 3\sqrt{2} + 15\sqrt{2} = 18\sqrt{2}$.

21. Since we must move the decimal point 7 places to the right to obtain 2.3,
    .000 000 23 $= 2.3 \times 10^{-7}$.

25. $\sqrt[4]{x^2+12} = \sqrt[4]{7x} \;\Rightarrow\; \left(\sqrt[4]{x^2+12}\right)^4 = \left(\sqrt[4]{7x}\right)^4 \;\Rightarrow\; x^2 + 12 = 7x \;\Rightarrow\;$
    $x^2 - 7x + 12 = 0 \;\Rightarrow\; (x-4)(x-3) \;\Rightarrow\; x = 3 \text{ or } 4$.

    Check $x = 3$: $\sqrt[4]{9+12} = \sqrt[4]{21}$ Check $x = 4$: $\sqrt[4]{16+12} = \sqrt[4]{28}$
                   $\sqrt[4]{21} = \sqrt[4]{21}$ ✓ $\sqrt[4]{28} = \sqrt[4]{28}$ ✓

    Therefore $x = 3$ and $x = 4$ are solutions.

29. Let x be the number. Then $\sqrt{x+1} = x - 5 \;\Rightarrow\; \left(\sqrt{x+1}\right)^2 = (x-5)^2 \;\Rightarrow\;$
    $x + 1 = x^2 - 10x + 25 \;\Rightarrow\; 0 = x^2 - 11x + 24 \;\Rightarrow\; 0 = (x-3)(x-8) \;\Rightarrow\; x = 3 \text{ or } 8$.
    Check $x = 3$: $\sqrt{3+1} = 3 - 5$ Check $x = 8$: $\sqrt{8+1} = 8 - 5$
                      $\sqrt{4} = -2$ $\sqrt{9} = 3$
                      $2 = -2$ ✗ $3 = 3$ ✓

    Therefore 8 is the only solution.

Chapter 8 Review Exercises and Test

33. $(\sqrt{2})^2 = 2, (\sqrt{7})^2 = 7$, and $3^2 = 9$. Since $2 + 7 = 9$, the numbers are the lengths of the sides of a right triangle.

37. Since $y = kx$, $-4 = 20k \Rightarrow k = \frac{-4}{20} = -\frac{1}{5}$. Therefore $y = -\frac{1}{5}x$. When $y = -8$
$-8 = -\frac{1}{5}x \Rightarrow 8 = \frac{1}{5}x \Rightarrow x = 5 \cdot 8 = 40$.

Chapter Test

1. $x^6(4x^{-2})^2 = x^6 \cdot 4^2 \cdot x^{-4} = 16x^6 \cdot x^{-4} = 16x^{6-4} = 16x^2$.

5. $\sqrt[3]{-16x^4} = \sqrt[3]{-8 \cdot 2 \cdot x^3 \cdot x} = \sqrt[3]{-8x^3} \cdot \sqrt[3]{2x} = -2x \cdot \sqrt[3]{2x}$.

9. $\frac{k^2 - 81}{\sqrt{k} - 3} = \frac{k^2 - 81}{\sqrt{k} - 3} \cdot \frac{\sqrt{k} + 3}{\sqrt{k} + 3} = \frac{(k+9)(k-9)(\sqrt{k} + 3)}{k - 9} = (k+9)(\sqrt{k} + 3)$.

13. $\frac{1}{\sqrt[3]{50}} = \frac{1}{\sqrt[3]{5^2 \cdot 2}} \cdot \frac{\sqrt[3]{5 \cdot 2^2}}{\sqrt[3]{5 \cdot 2^2}} = \frac{\sqrt[3]{5 \cdot 2^2}}{\sqrt[3]{5^3 \cdot 2^3}} = \frac{\sqrt[3]{5 \times 4}}{5 \times 2} = \frac{\sqrt[3]{20}}{10}$.

17. $x = 2 + \sqrt{x} \Rightarrow x - 2 = \sqrt{x} \Rightarrow x^2 - 4x + 4 = x \Rightarrow x^2 - 5x + 4 = 0 \Rightarrow$
$(x - 4)(x - 1) = 0 \Rightarrow x = 4$ or 1.
Check $x = 4$: $\quad 4 = 2 + \sqrt{4}$ $\qquad$ Check $x = 1$: $\quad 1 = 2 + \sqrt{1}$
$\qquad\qquad\qquad 4 = 2 + 2$ ✓ $\qquad\qquad\qquad\qquad\qquad 1 = 2 + 1$ X
Therefore 4 is the only solution.

21. $6^2 = 36, (3\sqrt{5})^2 = 9 \times 5 = 45$, and $9^2 = 81$. Since $36 + 45 = 81$ is true, there is a right triangle with sides $6, 3\sqrt{5}$, and 9.

25. If h is the height and b the base then $hb = k$. Since $b = 8$ when $h = 3$,
$8 \times 3 = k \Rightarrow k = 24$. Thus we know $hb = 24$. Now if the height of another triangle is 10,
$10b = 24 \Rightarrow b = \frac{24}{10} = \frac{12}{5}$.

Chapter 9: Quadratic Equations

Exercises 9.1

1. To solve $x^2 + 2x - 24 = 0$, factor the left side of the equation to obtain $(x+6)(x-4) = 0$. Then $x + 6 = 0$ or $x - 4 = 0$ which implies that $x = -6$ or 4.

5. $x^2 + x = 0 \Rightarrow x(x+1) = 0 \Rightarrow x = 0$ or $x + 1 = 0 \Rightarrow x = 0$ or -1.

9. To solve $\frac{x^2}{12} + \frac{2}{3} = \frac{x}{2}$ first multiply both sides by the LCD 12:
$12 \cdot \left(\frac{x^2}{12} + \frac{2}{3}\right) = 12 \cdot \frac{x}{2} \Rightarrow x^2 + 8 = 6x \Rightarrow x^2 - 6x + 8 = 0 \Rightarrow (x-4)(x-2) = 0 \Rightarrow$
$x - 4 = 0$ or $x - 2 = 0 \Rightarrow x = 4$ or 2.

13. To solve $\sqrt[3]{x^2 + 2x} = \sqrt[3]{15}$ cube both sides of the equation. This gives
$x^2 + 2x = 15 \Rightarrow x^2 + 2x - 15 = 0 \Rightarrow (x+5)(x-3) = 0 \Rightarrow x = -5$ or 3.

17. $x^2 = 25 \Rightarrow x = \pm\sqrt{25} \Rightarrow x = \pm 5$.

21. $2k^2 = 16 \Rightarrow k^2 = 8 \Rightarrow k = \pm\sqrt{8} \Rightarrow k = \sqrt{4 \cdot 2} \Rightarrow k = \pm 2\sqrt{2}$.

25. $2s^2 - 7 = 0 \Rightarrow 2s^2 = 7 \Rightarrow s^2 = \frac{7}{2} \Rightarrow s = \pm\sqrt{\frac{7}{2}} \Rightarrow s = \pm\frac{\sqrt{7}}{\sqrt{2}} \cdot \frac{\sqrt{2}}{\sqrt{2}} \Rightarrow$
$s = \pm\frac{\sqrt{14}}{\sqrt{4}} \Rightarrow s = \pm\frac{\sqrt{14}}{2}$.

29. $(k+1)^2 = 16 \Rightarrow k + 1 = \pm\sqrt{16} \Rightarrow k + 1 = \pm 4 \Rightarrow k = -4 - 1$ or $k = 4 - 1 \Rightarrow$
$k = -5$ or 3.

33. $\left(x - \frac{1}{2}\right)^2 = \frac{9}{4} \Rightarrow x - \frac{1}{2} = \pm\sqrt{\frac{9}{4}} \Rightarrow x - \frac{1}{2} = \pm\frac{3}{2} \Rightarrow x = \frac{1}{2} + \frac{3}{2}$ or $x = \frac{1}{2} - \frac{3}{2} \Rightarrow$
$x = \frac{4}{2}$ or $x = -\frac{2}{2} \Rightarrow x = 2$ or -1.

37. $d = 16t^2 \Rightarrow t^2 = \frac{d}{16} \Rightarrow t = \pm\sqrt{\frac{d}{16}} \Rightarrow t = \pm\frac{\sqrt{d}}{\sqrt{16}} \Rightarrow t = \pm\frac{\sqrt{d}}{4}$.

41 To simplify the equation $\frac{8T^2}{\pi^2} - L = 0$ multiply both sides by π^2. This gives

$8T^2 - \pi^2 L = 0 \Rightarrow 8T^2 = \pi^2 L \Rightarrow T^2 = \frac{\pi^2 L}{8} \Rightarrow T = \pm\sqrt{\frac{\pi^2 L}{8}} \Rightarrow$

$T = \pm\frac{\sqrt{\pi^2 L}}{\sqrt{8}} \cdot \frac{\sqrt{2}}{\sqrt{2}} \Rightarrow T = \pm\frac{\sqrt{\pi^2} \cdot \sqrt{2L}}{\sqrt{16}} \Rightarrow T = \pm\frac{\pi\sqrt{2L}}{4}$.

45 To solve the equation $d = \sqrt{L^2 + W^2}$ for L, square both sides. This gives

$d^2 = L^2 + W^2 \Rightarrow d^2 - W^2 = L^2 \Rightarrow L^2 = d^2 - W^2 \Rightarrow L = \pm\sqrt{d^2 - W^2}$.

49 $h = \frac{r^2 - v^2 r}{v^2} \Rightarrow hv^2 = r^2 - v^2 r \Rightarrow hv^2 + v^2 r = r^2 \Rightarrow v^2(h + r) = r^2 \Rightarrow$

$v^2 = \frac{r^2}{h+r} \Rightarrow v = \pm\sqrt{\frac{r^2}{h+r}} \Rightarrow v = \pm\frac{\sqrt{r^2}}{\sqrt{h+r}} \cdot \frac{\sqrt{h+r}}{\sqrt{h+r}} \Rightarrow v = \pm\frac{r\sqrt{h+r}}{h+r}$.

53 $S = 2\pi r\sqrt{X^2 + Y^2} \Rightarrow S^2 = 4\pi^2 r^2 (X^2 + Y^2) \Rightarrow S^2 = 4\pi^2 r^2 X^2 + 4\pi^2 r^2 Y^2 \Rightarrow$

$S^2 - 4\pi^2 r^2 Y^2 = 4\pi^2 r^2 X^2 \Rightarrow X^2 = \frac{S^2 - 4\pi^2 r^2 Y^2}{4\pi^2 r^2} \Rightarrow X = \pm\sqrt{\frac{S^2 - 4\pi^2 r^2 Y^2}{4\pi^2 r^2}} \Rightarrow$

$X = \pm\frac{\sqrt{S^2 - 4\pi^2 r^2 Y^2}}{\sqrt{4\pi^2 r^2}} \Rightarrow X = \pm\frac{\sqrt{S^2 - 4\pi^2 r^2 Y^2}}{2\pi r}$.

57 $(2x - 3)^2 = 5 \Rightarrow 2x - 3 = \pm\sqrt{5} \Rightarrow 2x = 3 \pm \sqrt{5} \Rightarrow x = \frac{3 \pm \sqrt{5}}{2}$.

Check $x = \frac{3+\sqrt{5}}{2}$: $\left(2 \cdot \frac{3+\sqrt{5}}{2} - 3\right)^2 = 5$ Check $x = \frac{3-\sqrt{5}}{2}$: $\left(2 \cdot \frac{3-\sqrt{5}}{2} - 3\right)^2 = 5$

$(3 + \sqrt{5} - 3)^2 = 5$ $\qquad\qquad$ $(3 - \sqrt{5} - 3)^2 = 5$

$(\sqrt{5})^2 = 5$ $\qquad\qquad\qquad$ $(-\sqrt{5})^2 = 5$

$5 = 5$ ✓ $\qquad\qquad\qquad$ $(-1)^2(\sqrt{5})^2 = 5$ or $5 = 5$ ✓

61 $(2t + 6)^2 = 8 \Rightarrow 2t + 6 = \pm\sqrt{8} \Rightarrow 2t = -6 \pm \sqrt{8} \Rightarrow t = \frac{-6 \pm \sqrt{8}}{2} \Rightarrow$

$t = \frac{-6 \pm \sqrt{4 \cdot 2}}{2} \Rightarrow t = \frac{-6 \pm 2\sqrt{2}}{2} \Rightarrow t = \frac{2(-3 \pm \sqrt{2})}{2} \Rightarrow t = -3 \pm \sqrt{2}$.

65 Let x be the number. Then $(x+2)^2 = 36 \Rightarrow x + 2 = \pm 6 \Rightarrow x = -2 \pm 6 \Rightarrow x = -8$ or 4.

Exercises 9.1

Review Problems

69 Find m and n such that $mn = 9$ and $m+n = 6$. Select $m = n = 3$. Then
$$x^2 + 6x + 9 = x^2 + 3x + 3x + 9 = x(x+3) + 3(x+3) = (x+3)(x+3).$$

73 Find m and n such that $mn = \frac{1}{4}$ and $m+n = -1$. Select $m = n = -\frac{1}{2}$. Then
$$y^2 - y + \frac{1}{4} = \left(y - \frac{1}{2}\right)\left(y - \frac{1}{2}\right).$$

Exercises 9.2

1 To complete the square of $x^2 + 8x$ add $\left(\frac{8}{2}\right)^2 = 4^2 = 16$. Note that $x^2 + 8x + 16 = (x+4)^2$.

5 To complete the square of $z^2 - 4z$ add $\left(\frac{-4}{2}\right)^2 = (-2)^2 = 4$.

9 To complete the square of $r^2 - \frac{3}{5}r$ add the square of half of $\frac{3}{5}$. Note you can use $\frac{3}{5}$ rather than $-\frac{3}{5}$. So first take half of $\frac{3}{5}$ which is $\frac{1}{2} \cdot \frac{3}{5} = \frac{3}{10}$. Then $\left(\frac{3}{10}\right)^2 = \frac{9}{100}$.
Therefore to complete the square, add $\frac{9}{100}$.

13 $x^2 - 18x + \left(\frac{18}{2}\right)^2 = x^2 - 18x + 9^2 = x^2 - 18x + 81 = (x-9)^2$.

17 $y^2 + \frac{2}{3}y + \left(\frac{1}{2} \cdot \frac{2}{3}\right)^2 = y^2 + \frac{2}{3}y + \left(\frac{1}{3}\right)^2 = y^2 + \frac{2}{3}y + \frac{1}{9} = \left(y + \frac{1}{3}\right)^2$.

21 a) $y^2 + 6y = 0$
$y(y+6) = 0$
$y = 0$ or $y+6 = 0$
$y = 0$ or -6.

b) $y^2 + 6y = 0$
$y^2 + 6y + 9 = 9$
$(y+3)^2 = 9$
$y + 3 = \pm 3$
$y = -3 \pm 3 \Rightarrow y = 0$ or -6.

Exercises 9.2

25 $x^2 + 4x = 1$
$x^2 + 4x + 4 = 1 + 4$
$(x+2)^2 = 5$
$x + 2 = \pm\sqrt{5}$
$x = -2 \pm \sqrt{5}$.

29 $x^2 = 4x$
$x^2 - 4x = 0$
$x^2 - 4x + 4 = 0 + 4$
$(x-2)^2 = 4$
$x - 2 = \pm 2$
$x = 2 \pm 2 \quad \Rightarrow \quad x = 0 \text{ or } 4$.

33 $(x-4)(x-5) = 20$
$x^2 - 9x + 20 = 20$
$x^2 - 9x = 0$
$x^2 - 9x + \left(\frac{9}{2}\right)^2 = 0 + \left(\frac{9}{2}\right)^2$
$\left(x - \frac{9}{2}\right)^2 = \frac{81}{4}$
$x - \frac{9}{2} = \pm\sqrt{\frac{81}{4}}$
$x = \frac{9}{2} \pm \frac{9}{2} \quad \Rightarrow \quad x = \frac{18}{2} \text{ or } 0 \quad \Rightarrow \quad x = 9 \text{ or } 0$.

37 To complete the square of $y^2 + \frac{2}{5}y$ we must add $\left(\frac{1}{2} \cdot \frac{2}{5}\right)^2 = \left(\frac{1}{5}\right)^2 = \frac{1}{25}$.
Therefore
$y^2 + \frac{2}{5}y = 0 \quad \Rightarrow$
$y^2 + \frac{2}{5}y + \frac{1}{25} = \frac{1}{25} \quad \Rightarrow$
$\left(y + \frac{1}{5}\right)^2 = \frac{1}{25} \quad \Rightarrow$
$y + \frac{1}{5} = \pm\sqrt{\frac{1}{25}} \quad \Rightarrow \quad y + \frac{1}{5} = \pm\frac{1}{5} \quad \Rightarrow$
$y = -\frac{1}{5} \pm \frac{1}{5} \quad \Rightarrow \quad y = 0 \text{ or } -\frac{2}{5}$.

41 a) $6z^2 = 4z$
$6z^2 - 4z = 0$
$2z(3z - 2) = 0$
$2z = 0 \text{ or } 3z - 2 = 0$
$z = 0 \text{ or } 3z = 2$
$z = 0 \text{ or } z = \frac{2}{3}$

b) $6z^2 - 4z = 0$
$\frac{1}{6} \cdot (6z^2 - 4z) = \frac{1}{6} \cdot 0$
$z^2 - \frac{2}{3}z = 0$
$z^2 - \frac{2}{3}z + \left(\frac{1}{3}\right)^2 = \left(\frac{1}{3}\right)^2$
$\left(z - \frac{1}{3}\right)^2 = \frac{1}{9}$
$x - \frac{1}{3} = \pm\sqrt{\frac{1}{9}} \quad \Rightarrow \quad x - \frac{1}{3} = \pm\frac{1}{3} \quad \Rightarrow$
$x = \frac{1}{3} \pm \frac{1}{3} \quad \Rightarrow \quad x = 0 \text{ or } \frac{2}{3}$.

45 $2y^2 + 4y - 3 = 0 \quad \Rightarrow \quad \frac{1}{2} \cdot (2y^2 + 4y - 3) = \frac{1}{2} \cdot 0 \quad \Rightarrow \quad y^2 + 2y - \frac{3}{2} = 0 \quad \Rightarrow$
$y^2 + 2y = \frac{3}{2} \quad \Rightarrow \quad y^2 + 2y + 1 = \frac{3}{2} + 1 \quad \Rightarrow \quad (y+1)^2 = \frac{5}{2} \quad \Rightarrow \quad y + 1 = \pm\sqrt{\frac{5}{2}} \quad \Rightarrow$
$y + 1 = \pm\frac{\sqrt{5}}{\sqrt{2}} \cdot \frac{\sqrt{2}}{\sqrt{2}} \quad \Rightarrow \quad y + \frac{2}{2} = \pm\frac{\sqrt{10}}{2} \quad \Rightarrow \quad y = -\frac{2}{2} \pm \frac{\sqrt{10}}{2} \quad \Rightarrow \quad y = \frac{-2 \pm \sqrt{10}}{2}$.

Exercises 9.2

49 $\frac{1}{2}y^2 + 3y = 2 \Rightarrow 2\cdot\left(\frac{1}{2}y^2 + 3y\right) = 2\cdot 2 \Rightarrow y^2 + 6y = 4 \Rightarrow y^2 + 6y + 9 = 4 + 9 \Rightarrow$
$(y+3)^2 = 13 \Rightarrow y+3 = \pm\sqrt{13} \Rightarrow y = -3 \pm \sqrt{13}$.

53 To solve $\frac{3t}{4} = \frac{4}{t} + \frac{5}{2}$ first get rid of the fractions by multiplying both sides by the LCD $4t$:

$4t \cdot \frac{3t}{4} = 4t \cdot \left(\frac{4}{t} + \frac{5}{2}\right) \Rightarrow 4t \cdot \frac{3t}{4} = 4t \cdot \frac{4}{t} + 4t \cdot \frac{5}{2} \Rightarrow 3t^2 = 16 + 10t$. Now multiply both sides by $\frac{1}{3}$ to make the coefficient of t^2 equal to 1:

$\frac{1}{3} \cdot 3t^2 = \frac{1}{3} \cdot (16 + 10t) \Rightarrow t^2 = \frac{16}{3} + \frac{10}{3}t \Rightarrow t^2 - \frac{10}{3}t = \frac{16}{3}$. To complete the square add $\left(\frac{1}{2} \cdot \frac{10}{3}\right)^2 = \left(\frac{5}{3}\right)^2 = \frac{25}{9}$ to both sides of the equation:

$t^2 - \frac{10}{3}t + \frac{25}{9} = \frac{16}{3} + \frac{25}{9} \Rightarrow \left(t - \frac{5}{3}\right)^2 = \frac{48}{9} + \frac{25}{9} \Rightarrow \left(t - \frac{5}{3}\right)^2 = \frac{73}{9} \Rightarrow$

$t - \frac{5}{3} = \pm\sqrt{\frac{73}{9}} \Rightarrow t = \frac{5}{3} \pm \frac{\sqrt{73}}{3} \Rightarrow t = \frac{5 \pm \sqrt{73}}{3}$.

57 Replace x by $-7 + \sqrt{3}$ in the equation $x^2 + 14x + 46 = 0$:

$(-7 + \sqrt{3})^2 + 14(-7 + \sqrt{3}) + 46 = 0$

$49 - 14\sqrt{3} + 3 - 98 + 14\sqrt{3} + 46 = 0$

$49 - 98 + 3 + 46 = 0$

$0 = 0 \checkmark$. Therefore $-7 + \sqrt{3}$ is a solution.

Review Problems

61 $3x^2 = 4x - 5$ is equivalent to $3x^2 - 4x + 5 = 0$. Therefore $a = 3$, $b = -4$, and $c = 5$.

65 To solve $\begin{array}{l} 6x + 2y = 6 \\ 4x - y = 11 \end{array}$ by substitution, solve for y in the 2nd equation. This gives $y = 4x - 11$.

Then substitute $4x - 11$ for y in the 1st equation:

$6x + 2(4x - 11) = 6 \Rightarrow 6x + 8x - 22 = 6 \Rightarrow 14x = 28 \Rightarrow x = 2$.

Then $y = 4 \cdot 2 - 11 = 8 - 11 = -3$. Therefore the solution is $x = 2$ and $y = -3$.

Exercises 9.3

1 If $x^2 + 9x + 10 = 0$ then $a = 1$, $b = 9$, and $c = 10$. Therefore the solutions are

$$x = \frac{-b \pm \sqrt{b^2 - 4ac}}{2a} = \frac{-9 \pm \sqrt{9^2 - 4 \cdot 1 \cdot 10}}{2 \cdot 1} = \frac{-9 \pm \sqrt{81 - 40}}{2} = \frac{-9 \pm \sqrt{41}}{2}.$$

5 If $x^2 + 6x + 9 = 0$ then $a = 1, b = 6$, and $c = 9$. Then $x = \frac{-6 \pm \sqrt{6^2 - 4 \cdot 1 \cdot 9}}{2 \cdot 1} = \frac{-6 \pm \sqrt{36 - 36}}{2} = \frac{-6 \pm 0}{2} = \frac{-6}{2} = -3$. Note that this quadratic equation has only one solution.

9 To solve $x^2 - 3x = 0$ using the quadratic formula, note that $a = 1$, $b = -3$, and $c = 0$.

Then $x = \frac{-(-3) \pm \sqrt{(-3)^2 - 4(1)(0)}}{2} = \frac{3 \pm \sqrt{9 - 0}}{2} = \frac{3 \pm \sqrt{9}}{2} = \frac{3 \pm 3}{2} = 3$ or 0.

13 To solve the quadratic equation $\frac{1}{2}x^2 + x = \frac{3}{4}$ by the quadratic formula we must put all the terms on one side of the equation. Therefore move the $\frac{3}{4}$ to the left side of the equation giving $\frac{1}{2}x^2 + x - \frac{3}{4} = 0$. We can make the coefficients integers by multiplying both sides of the equation by 4: $4 \cdot \left(\frac{1}{2}x^2 + x - \frac{3}{4}\right) = 4 \cdot 0 \Rightarrow 2x^2 + 4x - 3 = 0$. Now $a = 2$, $b = 4$, and $c = -3$

and $x = \frac{-4 \pm \sqrt{4^2 - 4(2)(-3)}}{2 \cdot 2} = \frac{-4 \pm \sqrt{16 - (-24)}}{4} = \frac{-4 \pm \sqrt{16 + 24}}{4} = \frac{-4 \pm \sqrt{40}}{4} = \frac{-4 \pm \sqrt{4 \cdot 10}}{4} = \frac{-4 \pm 2\sqrt{10}}{4} = \frac{2(-2 \pm \sqrt{10})}{4} = \frac{-2 \pm \sqrt{10}}{2}.$

17 a) $x^2 + 5x + 6 = 0 \Rightarrow (x + 3)(x + 2) = 0 \Rightarrow x + 3 = 0$ or $x + 2 = 0 \Rightarrow x = -3$ or -2.

b) $x^2 + 5x + 6 = 0 \Rightarrow x^2 + 5x = -6 \Rightarrow x^2 + 5x + \frac{25}{4} = -6 + \frac{25}{4} \Rightarrow$

$\left(x + \frac{5}{2}\right)^2 = -\frac{24}{4} + \frac{25}{4} \Rightarrow \left(x + \frac{5}{2}\right)^2 = \frac{1}{4} \Rightarrow x + \frac{5}{2} = \pm \frac{1}{2} \Rightarrow x = -\frac{5}{2} \pm \frac{1}{2} \Rightarrow$

$x = -\frac{6}{2}$ or $-\frac{4}{2} \Rightarrow x = -3$ or -2.

Exercises 9.3

c) If $x^2 + 5x + 6 = 0$ then $a = 1$, $b = 5$, and $c = 6$. Therefore

$$x = \frac{-5 \pm \sqrt{25 - 4 \cdot 1 \cdot 6}}{2 \cdot 1} = \frac{-5 \pm \sqrt{25 - 24}}{2} = \frac{-5 \pm 1}{2} = -3 \text{ or } -2.$$

21 $x^2 - 11x + 24 = 0 \Rightarrow (x-8)(x-3) = 0 \Rightarrow x = 8 \text{ or } 3.$

25 $5y^2 - 7 = 0 \Rightarrow 5y^2 = 7 \Rightarrow y^2 = \frac{7}{5} \Rightarrow y = \pm\sqrt{\frac{7}{5}} \Rightarrow y = \pm\frac{\sqrt{7}}{\sqrt{5}} \cdot \frac{\sqrt{5}}{\sqrt{5}} = \pm\frac{\sqrt{35}}{5}.$

29 $12x^2 + 12x = 0 \Rightarrow 12x(x+1) = 0 \Rightarrow 12x = 0 \text{ or } x+1 = 0 \Rightarrow x = 0 \text{ or } -1.$

33 $2x(x-4) = -3 \Rightarrow 2x^2 - 8x = -3 \Rightarrow 2x^2 - 8x + 3 = 0$. Therefore

$$x = \frac{-(-8) \pm \sqrt{(-8)^2 - 4 \cdot 2 \cdot 3}}{2 \cdot 2} = \frac{8 \pm \sqrt{64 - 24}}{4} = \frac{8 \pm \sqrt{40}}{4} = \frac{8 \pm 2\sqrt{10}}{4} =$$

$$\frac{2(4 \pm \sqrt{10})}{4} = \frac{4 \pm \sqrt{10}}{2}.$$

37 $\frac{x}{4} = \frac{x-2}{x} \Rightarrow x^2 = 4x - 8 \Rightarrow x^2 - 4x + 8 = 0 \Rightarrow x = \frac{4 \pm \sqrt{16 - 32}}{2} = \frac{4 \pm \sqrt{-16}}{2}.$

Since $\sqrt{-16}$ is not a real number, there are no real solutions.

41 If $4t^2 - 10t + 5 = 0$ then $t = \frac{10 \pm \sqrt{100 - 4 \cdot 4 \cdot 5}}{8} \Rightarrow t = \frac{10 \pm \sqrt{20}}{8} \Rightarrow$

$t = \frac{10 \pm 2\sqrt{5}}{8} \Rightarrow t = \frac{2(5 \pm \sqrt{5})}{8} \Rightarrow t = \frac{5 \pm \sqrt{5}}{4}.$

45 If the height is 25 feet then $25 = 40t - 16t^2 \Rightarrow 16t^2 - 40t + 25 = 0 \Rightarrow$
$(4t - 5)(4t - 5) = 0 \Rightarrow 4t - 5 = 0 \Rightarrow t = \frac{5}{4}$ sec.

Exercises 9.3

49 If $L = 10$ then $100 - 10W - W^2 = 0 \Rightarrow 0 = W^2 + 10W - 100 \Rightarrow$

$W = \dfrac{-10 \pm \sqrt{(10)^2 - 4 \cdot 1 \cdot (-100)}}{2 \cdot 1} = \dfrac{-10 \pm \sqrt{100 - (-400)}}{2} = \dfrac{-10 \pm \sqrt{100 + 400}}{2} =$

$\dfrac{-10 \pm \sqrt{500}}{2} = \dfrac{-10 \pm 10\sqrt{5}}{2} = \dfrac{2(-5 \pm 5\sqrt{5})}{2} = -5 \pm 5\sqrt{5}$.

Since W can't be negative, W must be $-5 + 5\sqrt{5}$.

53 Problem Statement: The length of a rectangle is twice its width, and its area is 4 square feet. Find the rectangle's width and length.

Let x be the width and $2x$ the length. Then $x(2x) = 4 \Rightarrow 2x^2 = 4 \Rightarrow x^2 = 2 \Rightarrow x = \sqrt{2}$. Thus the width is $\sqrt{2}$ ft and the length is $2\sqrt{2}$ ft .

57 Problem Statement: Two numbers differ by 2 and their product is 26. Find the numbers.

Let x and $x + 2$ be the two numbers. Then $x(x + 2) = 26 \Rightarrow x^2 + 2x = 26 \Rightarrow$
$x^2 + 2x + 1 = 26 + 1 \Rightarrow (x + 1)^2 = 27 \Rightarrow x + 1 = \pm\sqrt{27} \Rightarrow x = -1 \pm 3\sqrt{3}$.
If $x = -1 - 3\sqrt{3}$ then $x + 2 = -1 - 3\sqrt{3} + 2 = 1 - 3\sqrt{3}$.
If $x = -1 + 3\sqrt{3}$ then $x + 2 = -1 + 3\sqrt{3} + 2 = 1 + 3\sqrt{3}$.
Therefore the numbers can be $-1 - 3\sqrt{3}$ and $1 - 3\sqrt{3}$ or $-1 + 3\sqrt{3}$ and $1 + 3\sqrt{3}$.

61 Problem Statement: Mirna cycled 9 miles. Numerically, her rate was 3 times her time. Find her rate and time. Give the exact answer and a 1 decimal place approximation.

Let x be her time. Then $3x$ is her rate. Since $D = RT$, $3x(x) = 9 \Rightarrow 3x^2 = 9 \Rightarrow$
$x^2 = 3 \Rightarrow x = \sqrt{3}$. Therefore her time is $\sqrt{3}$ or 1.7 and her rate is $3\sqrt{3}$ or 5.2 .
Note that $5.2 \neq 3 \times 1.7$.

65 Problem Statement: The length of a rectangle is 3 cm more than the width. If the length is doubled and the width is increased by 1, then the area is increased by 90 cm^2. Find the length and width of the original rectangle.

Let w be the width and $w + 3$ the length. Then the bigger rectangle has length $2(w + 3)$ and width $w + 1$. Since its area is 90 more than the area of the smaller rectangle,
$2(w + 3)(w + 1) = (w + 3)w + 90 \Rightarrow 2(w^2 + 4w + 3) = w^2 + 3w + 90 \Rightarrow$

Exercises 9.3

$2w^2 + 8w + 6 = w^2 + 3w + 90 \Rightarrow w^2 + 5w - 84 = 0 \Rightarrow (w-7)(w+12) = 0 \Rightarrow w = 7$.

Hence the original rectangle has width 7 cm and length 10 cm.

[69] $2x^3 = 6x^2 + 10x \Rightarrow 2x^3 - 6x^2 - 10x = 0 \Rightarrow 2x(x^2 - 3x - 5) = 0 \Rightarrow$

$2x = 0$ or $x^2 - 3x - 5 = 0$. Then $x = 0$ or $x = \dfrac{3 \pm \sqrt{9 - 4 \cdot 1(-5)}}{2} = \dfrac{3 \pm \sqrt{29}}{2}$.

Therefore this equation has 3 solutions: 0, $\dfrac{3 + \sqrt{29}}{2}$, and $\dfrac{3 - \sqrt{29}}{2}$.

[73] $\sqrt{6}\,y^2 - \sqrt{2} = y \Rightarrow \sqrt{6}\,y^2 - y - \sqrt{2} = 0$. Then use the quadratic formula where
$a = \sqrt{6}$, $b = -1$, and $c = -\sqrt{2}$. Therefore

$y = \dfrac{-(-1) \pm \sqrt{(-1)^2 - 4\sqrt{6} \cdot (-\sqrt{2})}}{2\sqrt{6}} = \dfrac{1 \pm \sqrt{1 + 4\sqrt{12}}}{2\sqrt{6}} = \dfrac{1 \pm \sqrt{1 + 4\sqrt{4 \cdot 3}}}{2\sqrt{6}} =$

$\dfrac{1 \pm \sqrt{1 + 8\sqrt{3}}}{2\sqrt{6}} \cdot \dfrac{\sqrt{6}}{\sqrt{6}} = \dfrac{1 \cdot \sqrt{6} \pm \sqrt{1 + 8\sqrt{3}} \cdot \sqrt{6}}{2 \cdot 6} = \dfrac{\sqrt{6} \pm \sqrt{(1 + 8\sqrt{3}) \cdot 6}}{12} =$

$\dfrac{\sqrt{6} \pm \sqrt{6 + 48\sqrt{3}}}{12}$.

Review Problems

[77] To find the x-intercept of the line $y = 4x - 1$, make $y = 0$ and solve for x:

$0 = 4x - 1 \Rightarrow 1 = 4x \Rightarrow x = \frac{1}{4}$. Therefore the x-intercept is $\frac{1}{4}$.

To find the y-intercept, make $x = 0$:

$y = 4 \cdot 0 - 1 = -1$. Therefore the y-intercept is -1.

Exercises 9.4

1. The parabolas given by $y = x^2 + 5$ and $y = x^2$ are identical in size and shape. However the graph of $y = x^2 + 5$ lies 5 units above the graph of $y = x^2$.

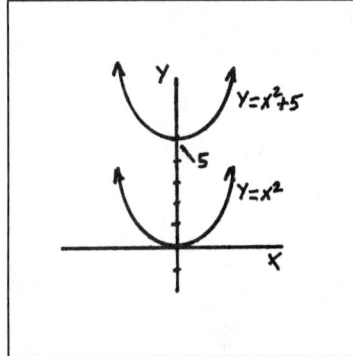

Figure 1

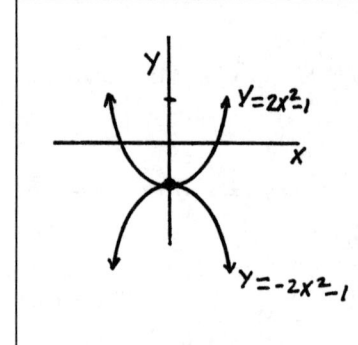

Figure 5

5. The parabolas given by $y = 2x^2 - 1$ and $y = -2x^2 - 1$ are identical in size and shape and each has vertex $(0, -1)$. However $y = 2x^2 - 1$ opens upward and $y = -2x^2 - 1$ opens downward.

9. Some points on the graph of $y = \frac{5}{4}x^2$ are $(0, 0)$, $(\pm 1, \frac{5}{4})$, and $(\pm 2, 5)$.

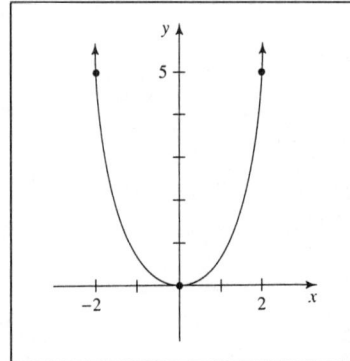

Figure 9

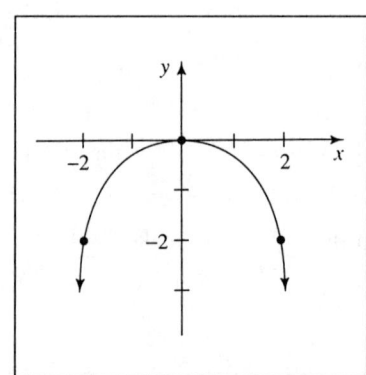

Figure 13

13. If $x = 0$ the $y = -\frac{1}{2} \cdot 0^2 = 0$. Therefore $(0, 0)$ is a point on the graph. If $x = -1$ then $y = -\frac{1}{2} \cdot (-1)^2 = -\frac{1}{2} \cdot 1 = -\frac{1}{2}$. Therefore $(-1, -\frac{1}{2})$ lies on the graph. Similarly $(1, -\frac{1}{2})$, $(2, -2)$, and $(-2, -2)$ lie on the graph. Note that the graph is a parabola opening downward.

Exercises 9.4

17 Some points on the graph of $y = -\frac{3}{4}x^2$ are $(0, 0)$, $(\pm 1, -\frac{3}{4})$, and $(\pm 2, -3)$. The graph is a parabola opening downward.

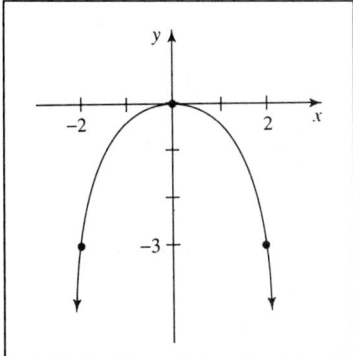

Figure 17

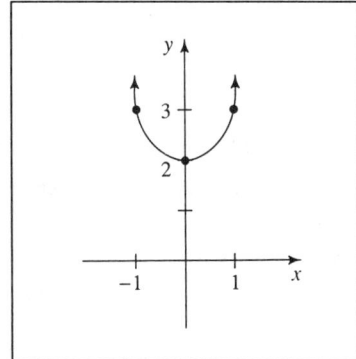

Figure 21

21 The vertex of the parabola $y = x^2 + 2$ is $(0, 2)$. Other points on the graph are $(\pm 1, 3)$ and $(\pm 2, 6)$. To find the roots, let $y = 0$ and solve for x: $0 = x^2 + 2 \Rightarrow x^2 = -2$. Therefore there are no real roots and consequently there are no x-intercepts.

25 The parabola given by the equation $y = -4x^2 - 1$ has vertex $(0, -1)$. This parabola passes through the points $(\pm 1, -5)$ and $(\pm 2, -17)$. It opens downward since the coefficient of x^2 is negative. To find the roots solve $0 = -4x^2 - 1 \Rightarrow 4x^2 = -1 \Rightarrow x^2 = -\frac{1}{4}$ which means there are no real roots and no x-intercepts.

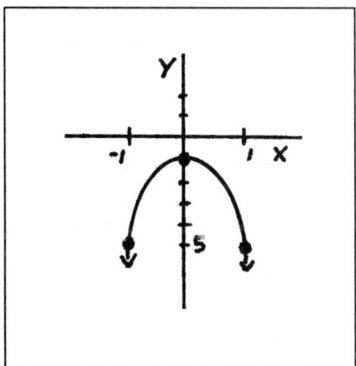

Figure 25

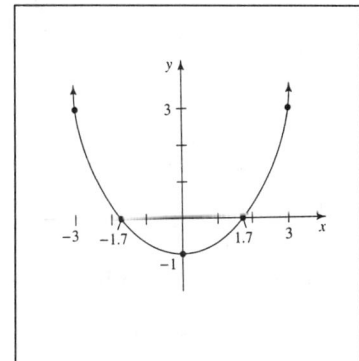

Figure 29

29 The vertex of the parabola $y = \frac{1}{3}x^2 - 1$ is $(0, -1)$. Some other points on this graph are $(\pm 1, -\frac{2}{3})$, and $(\pm 3, 2)$. To find the roots and x-intercepts let $y = 0$. Then $0 = \frac{1}{3}x^2 - 1 \Rightarrow 1 = \frac{1}{3}x^2 \Rightarrow x^2 = 3 \Rightarrow x = \pm\sqrt{3} \approx \pm 1.7$.

33 The parabola defined by $y = x^2 + \frac{1}{2}$ has vertex $(0, \frac{1}{2})$. The parabola passes through the points $(\pm 1, \frac{3}{2})$ and $(\pm 2, \frac{9}{2})$. If $y = 0$ then $0 = x^2 + \frac{1}{2} \Rightarrow x^2 = -\frac{1}{2}$ and consequently there are no real roots, and the graph does not cross the x-axis.

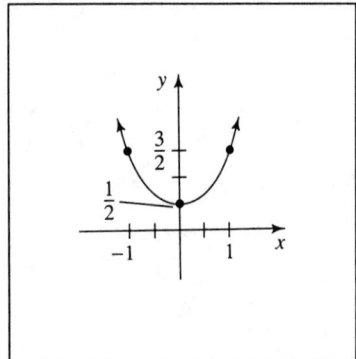
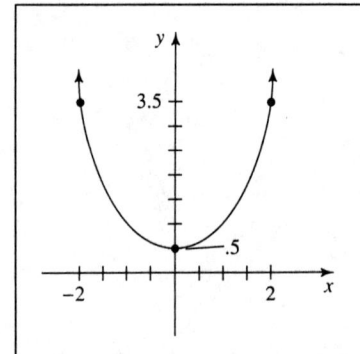

Figure 33   Figure 37

37 The vertex of the graph of $y = \frac{3}{4}x^2 + \frac{1}{2}$ is $(0, \frac{1}{2})$. Some other points on the graph are $(\pm 1, \frac{5}{4})$ and $(\pm 2, \frac{7}{2})$. If $y = 0$ then $0 = \frac{3}{4}x^2 + \frac{1}{2} \Rightarrow \frac{3}{4}x^2 = -\frac{1}{2} \Rightarrow x^2 = -\frac{2}{3}$ and there are no real roots and the graph has no x-intercepts.

41 The graph of $y = -x^2$ is a parabola opening downward with vertex $(0, 0)$. The parabola passes through the points $(\pm 1, -1)$ and $(\pm 2, -4)$.

$y = -1$ is a horizontal line with y-intercept -1.

From the graph we see that the solutions to the system $\begin{array}{l} y = -x^2 \\ y = -1 \end{array}$ are $(\pm 1, -1)$.

Page 190

Exercises 9.4

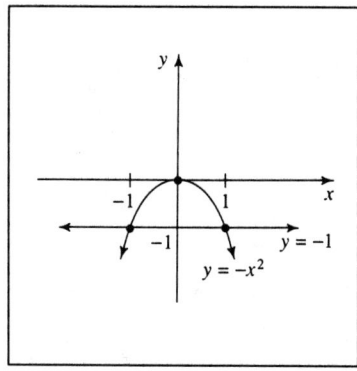

Figure 41

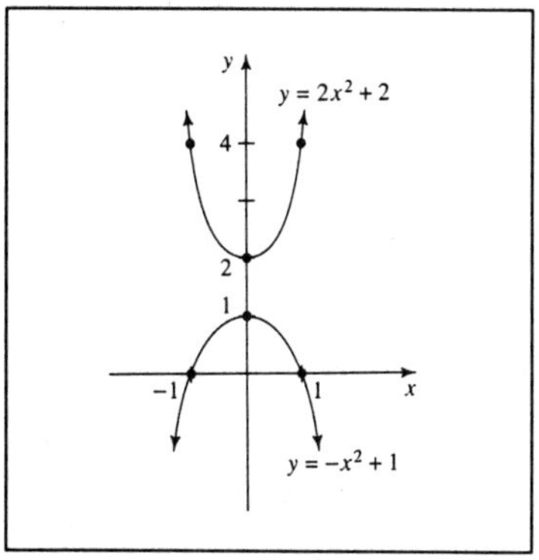

Figure 45

[45] $y = x^2 + 2$ is a parabola opening upward with vertex $(0, 2)$. $y = -x^2 + 1$ is a parabola opening downward with with vertex $(0, 1)$. There are no points of intersection, and consequently the system has no real solutions.

Review Problems

[49] To complete the square of $x^2 + 2x$ add $\left(\frac{2}{1}\right)^2 = 1^2 = 1$. Then $x^2 + 2x + 1 = (x+1)^2$.

[53] $-x^2 + 2x + 1 = 0 \;\Rightarrow\; x^2 - 2x - 1 = 0 \;\Rightarrow\; x^2 - 2x = 1 \;\Rightarrow\; x^2 - 2x + 1 = 1 + 1 \;\Rightarrow\;$
$(x-1)^2 = 2 \;\Rightarrow\; x - 1 = \pm\sqrt{2} \;\Rightarrow\; x = 1 \pm \sqrt{2}$.

Exercises 9.5

[1] The x-intercepts of $y = x^2 + 2x - 8$ are 2 and -4. To find the positive x-intercept using the graphing calculator, press RANGE and make Xmin $= 0$, Xmax $= 4$, Ymin $= -1$, and Ymax $= 1$. Press $\boxed{Y=}$ and make $y_1 = x^2 + 2x - 8$. Finally press TRACE and use the left and right arrow keys to place the trace bug on the point where the curve crosses the x-axis. To find the negative x-intercept, press RANGE and make Xmin $= -5$, Xmax $= -3$, Ymin $= -1$, and Ymax $= 1$. Then press TRACE and repeat the process.

Exercises 9.5

5) $y = x^2 + 6x + 9 = (x+3)^2$.

a) The vertex has x-coordinate -3 and $y = (-3+3)^2 = 0^2 = 0$. Therefore the vertex is $(-3, 0)$.

b) The y-intercept is the y value when $x = 0$. Therefore the y-intercept is $y = (0+3)^2 = 9$.

c) To find the roots, let $y = 0$: $0 = (x+3)^2 \Rightarrow x+3 = 0 \Rightarrow x = -3$. So -3 is the only root. Therefore the graph hits the x-axis at just one point $(-3, 0)$.

d) Some other points on the graph, in addition to the vertex $(-3, 0)$, are $(-2, 1)$, $(-4, 1)$, $(-1, 4)$, and $(-5, 4)$.

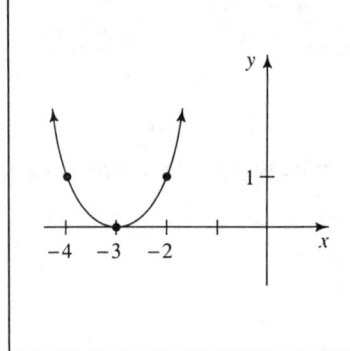

9) $y = x^2 - 4x + 5 \Rightarrow y = (x^2 - 4x + 4) - 4 + 5 \Rightarrow y = (x-2)^2 + 1$.

a) The vertex has x-coordinate 2. Then $y = (2-2)^2 + 1 = 0^2 + 1 = 1$. Therefore the vertex is $(2, 1)$.

b) If $x = 0$ then $y = 0^2 - 4 \cdot 0 + 5 = 5$. Therefore the y-intercept is 5.

c) $0 = (x-2)^2 + 1 \Rightarrow -1 = (x-2)^2 \Rightarrow x - 2 = \pm\sqrt{-1} \Rightarrow x = 2 \pm i$.

Therefore there are no real roots, and the graph does not cross the x-axis.

d) The graph passes through the points $(1, 2)$, $(3, 2)$, $(0, 5)$, and $(4, 5)$.

Exercises 9.5

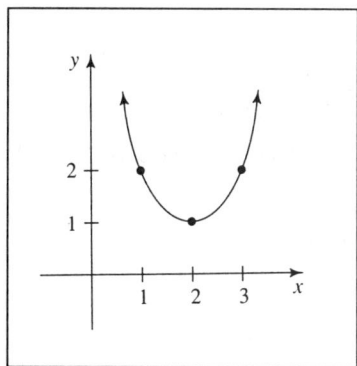

Figure 9

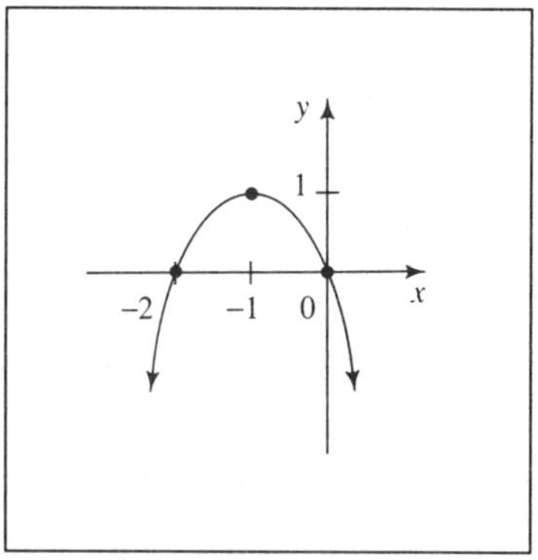

Figure 13

13 a) The vertex of $y = -(x+1)^2 + 1$ has x-coordinate -1. Then $y = -(-1+1)^2 + 1 = 1$ and the vertex is $(-1, 1)$.

b) The y-intercept is $y = -(0+1)^2 + 1 = -1^2 + 1 = -1 + 1 = 0$.

c) $0 = -(x+1)^2 + 1 \Rightarrow 1 = (x+1)^2 \Rightarrow x+1 = \pm 1 \Rightarrow x = -1 \pm 1 \Rightarrow$
$x = 0$ or -2. Therefore the roots are 0 and -2.

d) From parts a and d we know the graph has vertex $(-1, 1)$ and hits the x-axis at $(0, 0)$ and $(-2, 0)$. The graph also passes through $(1, -3)$ and $(-3, -3)$.

17 $y = -x^2 + 6x - 8 = -(x^2 - 6x) - 8 = -(x^2 - 6x + 9) + 9 - 8 = -(x-3)^2 + 1$.

a) If $x = 3$ then $y = 1$. Therefore the vertex is $(3, 1)$.

b) When $x = 0$ then $y = -8$. Therefore the y-intercept is -8.

c) $-(x-3)^2 + 1 = 0 \Rightarrow 1 = (x-3)^2 \Rightarrow x - 3 = \pm 1 \Rightarrow x = 3 \pm 1 \Rightarrow x = 4$ or 2.
Therefore the roots and x-intercepts are 2 and 4.

d) The graph has vertex $(3, 1)$ and crosses the x-axis at $(2, 0)$ and $(4, 0)$.

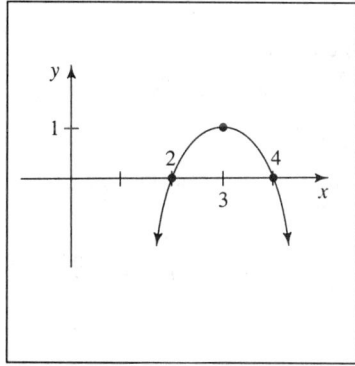

Figure 17

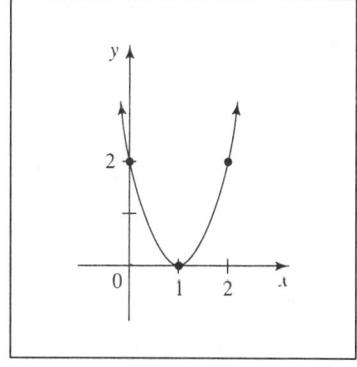

Figure 21

21. a) If $x = 1$, then $y = 2(1-1)^2 = 0$. Therefore the vertex is $(1, 0)$.
    b) If $x = 0$ then $y = 2(0-1)^2 = 2 \cdot 1 = 2$. Therefore the y-intercept is 2.
    c) If $y = 0$ then $0 = 2(x-1)^2 \Rightarrow (x-1)^2 = 0 \Rightarrow x = 1$. Therefore the only root is 1 and the x-intercept is 1.
    d) Some additional points on the graph are $(2, 2)$, $(-1, 8)$, and $(3, 8)$.

25. $y = 4x^2 - 8x + 3 = 4(x^2 - 2x) + 3 = 4(x^2 - 2x + 1) - 4 + 3 = 4(x-1)^2 - 1$.
    a) If $x = 1$ then $y = -1$ and the vertex is $(1, -1)$.
    b) If $x = 0$ then $y = 3$ and the y-intercept is 3.
    c) If $y = 0$ then $0 = 4(x-1)^2 - 1 \Rightarrow 1 = 4(x-1)^2 \Rightarrow (x-1)^2 = \frac{1}{4} \Rightarrow x - 1 = \pm\frac{1}{2} \Rightarrow$
       $x = 1 \pm \frac{1}{2} \Rightarrow x = 1.5$ or $.5$. Therefore the roots and x-intercepts are .5 and 1.5.
    d) To sketch the graph plot $(1, -1)$, $(0, 3)$, $(.5, 0)$, and $(1.5, 0)$. Also $(2, 3)$ is on the graph.

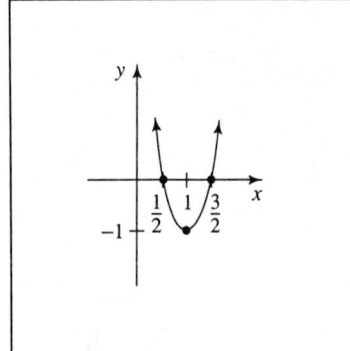

Figure 25

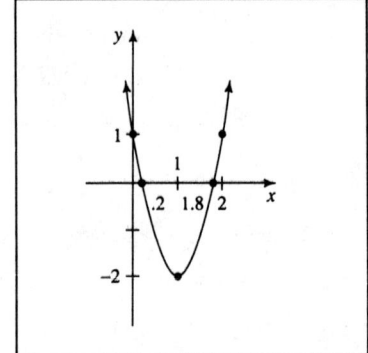

Figure 29

29. $y = 3x^2 - 6x + 1 = 3(x^2 - 2x) + 1 = 3(x^2 - 2x + 1) - 3 + 1 = 3(x-1)^2 - 2$.
    a) The vertex is $(1, -2)$.   b) The y-intercept is 1.
    c) $0 = 3(x-1)^2 - 2 \Rightarrow (x-1)^2 = \frac{2}{3} \Rightarrow x - 1 = \pm\sqrt{\frac{2}{3}} \Rightarrow x = 1 \pm \sqrt{\frac{2}{3}} = 1 \pm .8 \approx$
       1.8 or .2. Therefore the roots and x-intercepts are .2 and 1.8.
    d) In addition to $(1, -2)$, $(0, 1)$, $(.2, 0)$, and $(1.8, 0)$, you can plot the point $(2, 1)$.

33. $y = 1000x - x^2 = -x^2 + 1000x = -(x^2 - 1000x) = -\left(x^2 - 1000x + \left[\frac{1000}{2}\right]^2\right) + \left(\frac{1000}{2}\right)^2 =$
    $-(x-500)^2 + (500)^2 = -(x-500)^2 + 250{,}000$.
    Therefore the vertex is $(500, 250000)$.

[37] $y = -x^2 + 4x = -(x^2 - 4x) = -(x^2 - 4x + 4) + 4 = -(x-2)^2 + 4$.

Then the vertex is $(2, 4)$. If $y = 0$ then $-x^2 + 4x = 0 \Rightarrow -x(x-4) = 0 \Rightarrow x = 0$ or 4.
Therefore $(0, 0)$ and $(4, 0)$ lie on the graph. Some additional points on the graph are $(1, 3)$ and $(3, 3)$.

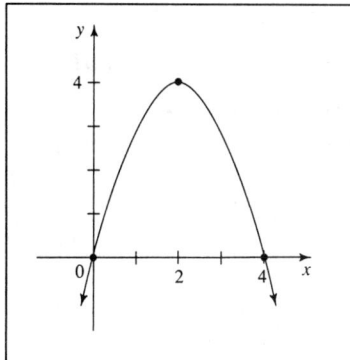

Figure 37

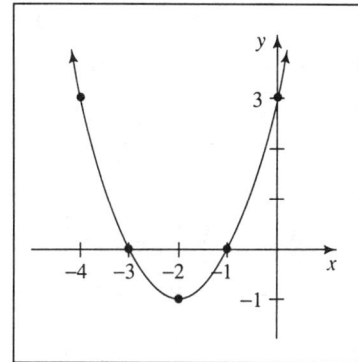

Figure 41

[41] $y = x^2 + 4x + 3 = (x^2 + 4x + 4) - 4 + 3 = (x+2)^2 - 1$ and we see the vertex is $(-2, -1)$.
$0 = x^2 + 4x + 3 \Rightarrow 0 = (x+3)(x+1) \Rightarrow x = -3$ or -1. Therefore $(-3, 0)$ and $(-1, 0)$ lie on the graph.

[45] $y = 2x^2 + 4x + 1 = 2(x^2 + 2x) + 1 = 2(x^2 + 2x + 1) - 2 + 1 = 2(x+1)^2 - 1$.

Thus the vertex is $(-1, -1)$. If $y = 0$ then $0 = 2(x+1)^2 - 1 \Rightarrow (x+1)^2 = \frac{1}{2} \Rightarrow$
$x + 1 = \pm\sqrt{\frac{1}{2}} \approx \pm.7 \Rightarrow x = -1 \pm .7 \Rightarrow x = -.3$ or -1.7. Therefore the roots are $-.3$ and -1.7. To graph the parabola, plot some additional points such as $(0, 1)$, $(-2, 1)$, $(1, 7)$, and $(-3, 7)$.

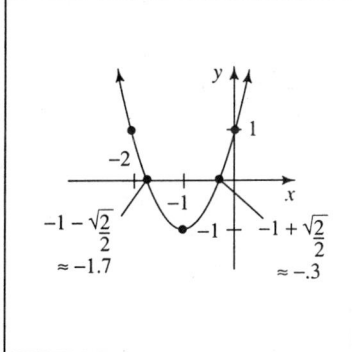

Figure 45

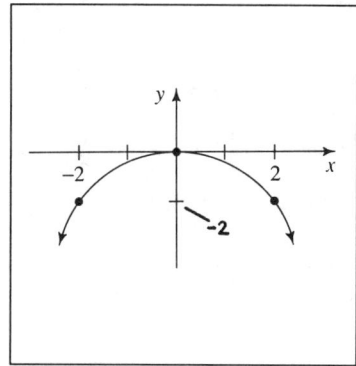

Figure 49

Exercises 9.5

49 The parabola given by $y = -\frac{1}{2}x^2$ opens downward with vertex $(0, 0)$. 0 is the only root. Some additional points on the graph are $(\pm 1, -\frac{1}{2})$ and $(\pm 2, -2)$.

53 $y = -3x^2 - 12x - 11 = -3(x^2 + 4x) - 11 = -3(x^2 + 4x + 4) + 12 - 11 = -3(x+2)^2 + 1$.
Therefore the vertex is $(-2, 1)$. If $y = 0$ then $-3(x+2)^2 + 1 = 0 \Rightarrow$
$\frac{1}{3} = (x+2)^2 \Rightarrow x + 2 = \pm\sqrt{\frac{1}{3}} \Rightarrow x = -2 \pm \sqrt{\frac{1}{3}} = -2 \pm .58 = -1.42$ or -2.58.
To sketch the graph, plot some additional points such as $(-3, -2)$ and $(-1, -2)$.

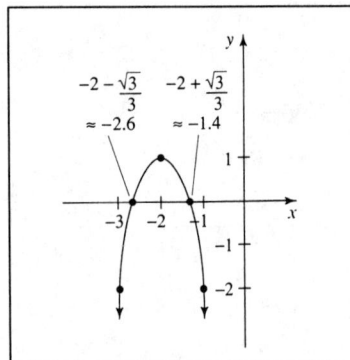

Figure 53

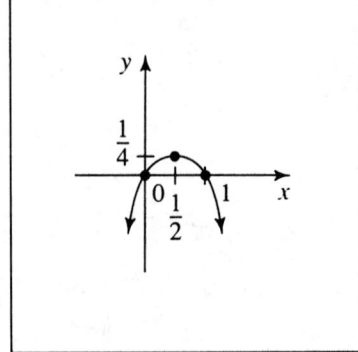

Figure 57

57 $y = x - x^2 \Rightarrow y = -x^2 + x \Rightarrow y = -(x^2 - x) \Rightarrow y = -\left(x^2 - x + \frac{1}{4}\right) + \frac{1}{4} \Rightarrow$
$y = -\left(x - \frac{1}{2}\right)^2 + \frac{1}{4}$. Therefore the vertex is $(\frac{1}{2}, \frac{1}{4})$. If $y = 0$ then $0 = x - x^2 \Rightarrow$
$0 = x(1 - x) \Rightarrow x = 0$ or 1. Therefore the roots are 0 and 1. Some other points on the graph in addition to $(\frac{1}{2}, \frac{1}{4})$, $(0, 0)$, and $(1, 0)$ are $(-1, -2)$ and $(2, -2)$.

61 $y = \frac{2}{3}x^2 - \frac{4}{3}x + \frac{2}{3} \Rightarrow y = \frac{2}{3}(x^2 - 2x + 1) \Rightarrow y = \frac{2}{3}(x - 1)^2$. Thus the vertex is $(1, 0)$.

65 $y = x - 3x^2 \Rightarrow y = -3x^2 + x \Rightarrow y = -3\left(x^2 - \frac{1}{3}x\right)$. To complete the square of $x^2 - \frac{1}{3}x$ add $\left(\frac{1}{2} \cdot \frac{1}{3}\right)^2 = \left(\frac{1}{6}\right)^2 = \frac{1}{36}$. Therefore
$y = -3\left(x^2 - \frac{1}{3}x + \frac{1}{36}\right) + 3 \cdot \frac{1}{36} \Rightarrow y = -3\left(x - \frac{1}{6}\right)^2 + \frac{1}{12}$.
Then we see that the vertex is $\left(\frac{1}{6}, \frac{1}{12}\right)$.

Chapter Review Exercises

1. $3x^2 + 2x = 0 \Rightarrow x(3x+2) = 0 \Rightarrow x = 0$ or $3x + 2 = 0 \Rightarrow x = 0$ or $-\frac{2}{3}$.

5. $3x^2 = 27 \Rightarrow x^2 = 9 \Rightarrow x = \pm 3$.

9. $x^2 + 4x - 2 = 0 \Rightarrow x^2 + 4x = 2 \Rightarrow x^2 + 4x + 4 = 2 + 4 \Rightarrow (x+2)^2 = 6 \Rightarrow$
   $x + 2 = \pm\sqrt{6} \Rightarrow x = -2 \pm \sqrt{6}$.

13. $3x^2 = 7x \Rightarrow 3x^2 - 7x = 0 \Rightarrow x(3x - 7) = 0 \Rightarrow x = 0$ or $3x - 7 = 0 \Rightarrow x = 0$ or $\frac{7}{3}$.

17. $(x - 5)^2 = 20 \Rightarrow x - 5 = \pm\sqrt{20} \Rightarrow x = 5 \pm \sqrt{20} \Rightarrow x = 5 \pm 2\sqrt{5}$.

21. Since $d = 16t^2$ and $d = 45$, $45 = 16t^2 \Rightarrow t^2 = \frac{45}{16} \Rightarrow t = \sqrt{\frac{45}{16}} = \frac{\sqrt{9 \cdot 5}}{\sqrt{16}} = \frac{3\sqrt{5}}{4}$ sec.

25. The parabola $y = -x^2 + 1$ opens downward with vertex $(0, 1)$. It has x-intercepts ± 1 and goes through the points $(2, -3)$ and $(-2, -3)$.

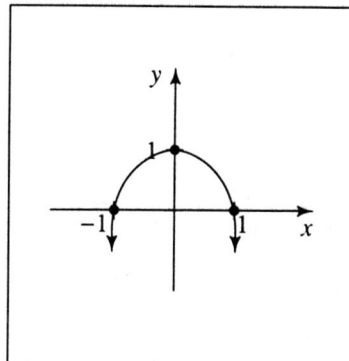

Figure 25

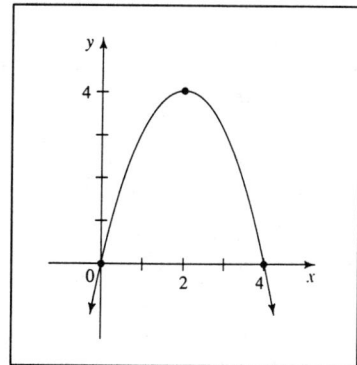

Figure 29

29. $y = -x^2 + 4x = -(x^2 - 4x) = -(x^2 - 4x + 4) + 4 = -(x - 2)^2 + 4$. Thus we see the vertex is $(2, 4)$. $0 = -x^2 + 4x \Rightarrow -x(x - 4) = 0 \Rightarrow x = 0$ or 4. Therefore the parabola crosses the x-axis at $(0, 0)$ and $(4, 0)$.

Chapter Test

1. $5x^2 - 4x = 0 \Rightarrow x(5x - 4) = 0 \Rightarrow x = 0$ or $5x - 4 = 0 \Rightarrow x = 0$ or $\frac{4}{5}$.

5. $2x^2 - 12x = -14 \Rightarrow x^2 - 6x = -7 \Rightarrow x^2 - 6x + 9 = -7 + 9 \Rightarrow (x - 3)^2 = 2 \Rightarrow$
   $x - 3 = \pm\sqrt{2} \Rightarrow x = 3 \pm \sqrt{2}$.

9. $x^2 = 17 \Rightarrow x = \pm\sqrt{17}$.

13. $x^2 + 2x = 100 \Rightarrow x^2 + 2x + 1 = 101 \Rightarrow (x + 1)^2 = 101 \Rightarrow x + 1 = \pm\sqrt{101} \Rightarrow$
    $x = -1 \pm \sqrt{101}$.

17. Let x be the 1st number and $8 - x$ the 2nd number. Then $x(8 - x) = 13 \Rightarrow 8x - x^2 = 13 \Rightarrow$
    $-13 = x^2 - 8x \Rightarrow x^2 - 8x = -13 \Rightarrow x^2 - 8x + 16 = -13 + 16 \Rightarrow$
    $(x - 4)^2 = 3 \Rightarrow x - 4 = \pm\sqrt{3} \Rightarrow x = 4 \pm \sqrt{3}$.
    If $x = 4 + \sqrt{3}$ then $8 - x = 8 - (4 + \sqrt{3}) = 8 - 4 - \sqrt{3} = 4 - \sqrt{3}$.
    If $x = 4 - \sqrt{3}$ then $8 - x = 8 - (4 - \sqrt{3}) = 8 - 4 + \sqrt{3} = 4 + \sqrt{3}$.
    In any case, the two numbers are $4 + \sqrt{3}$ and $4 - \sqrt{3}$.

21. $y = -\frac{1}{4}x^2 + 1$ has vertex $(0, 1)$. If $y = 0$ then $0 = -\frac{1}{4}x^2 + 1 \Rightarrow \frac{1}{4}x^2 = 1 \Rightarrow x^2 = 4 \Rightarrow$
    $x = \pm 2$. Therefore the x-intercepts are ± 2 and the graph crosses the x-axis at $(\pm 2, 0)$.

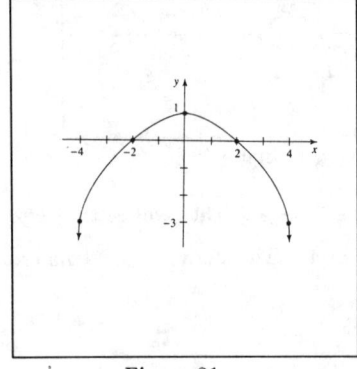

Figure 21

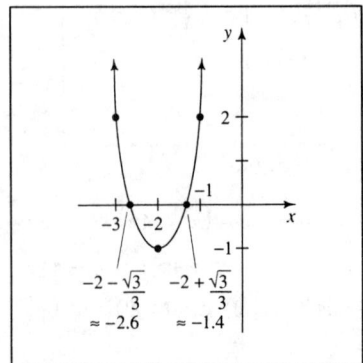

Figure 25

25. $y = 3x^2 + 12x + 11 \Rightarrow y = 3(x^2 + 4x) + 11 \Rightarrow y = 3(x^2 + 4x + 4) - 12 + 11 \Rightarrow$
$y = 3(x+2)^2 - 1$. Therefore the vertex of this parabola is $(-2, -1)$.
If $y = 0$ then $0 = 3(x+2)^2 - 1 \Rightarrow 1 = 3(x+2)^2 \Rightarrow \frac{1}{3} = (x+2)^2 \Rightarrow$
$x + 2 = \pm\sqrt{\frac{1}{3}} \Rightarrow x \approx -2 \pm .6 = -2.6$ or -1.4. Therefore the graph crosses the x-axis at approximately $(-2.6, 0)$ and $(-1.4, 0)$.

Chapter 10: Additional Topics

Exercises 10.1

1. $\sqrt{-4} = \sqrt{4}\,i = 2i$.

5. $\sqrt{-32} = \sqrt{32}\,i = \sqrt{16\cdot 2}\,i = \sqrt{16}\cdot\sqrt{2}\,i = 4\sqrt{2}\,i$.

9. $\sqrt[3]{-1} + \sqrt{-1} = -1 + i$.

13. $x^2 = -16 \Rightarrow x = \pm\sqrt{-16} = \pm\sqrt{16}\,i = \pm 4i$.
Check: $(4i)^2 = 16\,i^2 = 16\cdot(-1) = -16\ \checkmark;\ (-4i)^2 = (-4)^2 i^2 = 16\cdot(-1) = -16\ \checkmark.$

17. $x^2 + 1 = -3 \Rightarrow x^2 = -4 \Rightarrow x = \pm\sqrt{-4} \Rightarrow x = \pm 2i$.

21.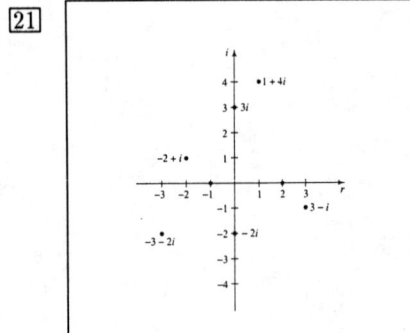

25. $(2+4i)+(6+8i) = (2+6)+(4i+8i) = 8+12i$.

29. $7+(6-8i) = (7+6)-8i = 13-8i$.

33. $5i - (-1+3i) + (4i-7) = 5i+1-3i+4i-7 = (1-7)+(5i-3i+4i) = -6+6i$.

37. $-4i(1-2i) = -4i\cdot 1 - 4i\cdot(-2i) = -4i+8i^2 = -4i+8\cdot(-1) = -4i-8 = -8-4i$.

41. $(3+4i)^2 = (3+4i)(3+4i) = 9+12i+12i+16i^2 = 9+24i+16\cdot(-1) = 9+24i-16 = (9-16)+24i = -7+24i$.

Exercises 10.1

[45] $\frac{1}{2+5i} = \frac{1}{2+5i} \cdot \frac{2-5i}{2-5i} = \frac{1 \cdot (2-5i)}{(2+5i)(2-5i)} = \frac{2-5i}{4-10i+10i-25i^2} = \frac{2-5i}{4-25 \cdot (-1)} =$
$\frac{2-5i}{4+25} = \frac{2-5i}{29} = \frac{2}{29} - \frac{5}{29}i$.

[49] $\frac{2+i}{3-i} = \frac{2+i}{3-i} \cdot \frac{3+i}{3+i} = \frac{(2+i)(3+i)}{(3-i)(3+i)} = \frac{6+2i+3i+i^2}{9+3i-3i-i^2} = \frac{6+5i+(-1)}{9-(-1)} = \frac{5+5i}{10} =$
$\frac{5}{10} + \frac{5}{10}i = \frac{1}{2} + \frac{1}{2}i$.

[53] $\frac{1+i}{2i} = \frac{1+i}{2i} \cdot \frac{-i}{-i} = \frac{-i(1+i)}{2i \cdot (-i)} = \frac{-i-i^2}{-2i^2} = \frac{-i-(-1)}{-2 \cdot (-1)} = \frac{-i+1}{2} = -\frac{i}{2} + \frac{1}{2} = \frac{1}{2} - \frac{1}{2}i$.

[57] $i^4 = (i^2)^2 = (-1)^2 = 1$. **[61]** $i^8 = (i^2)^4 = (-1)^4 = 1$.

[65] $\sqrt{16} = 4$ is a natural number, a whole number, an integer, a rational number, a real number, and a complex number. It is not an irrational number. It is not undefined.

[69] 1.343536... is an irrational number, a real number, and a complex number. It is not a natural number, a whole number, an integer, or a rational number.

[73] $3 + 2i$ is a complex number. It is not a natural number, a whole number, an integer, a rational number, an irrational number, or a real number.

[77] Note that $(3i)^2 = 9i^2 = 9 \cdot (-1) = -9$. Therefore, since $a^2 - b^2 = (a+b)(a-b)$, $a^2 + 9 = a^2 - (-9) = a^2 - (3i)^2 = (a+3i)(a-3i)$.

<u>Review Problems</u>

[85] The solutions to $3x^2 - 2x - 1 = 0$ are $x = \frac{2 \pm \sqrt{4 - 4(3)(-1)}}{6} = \frac{2 \pm \sqrt{4+12}}{6} =$
$\frac{2 \pm \sqrt{16}}{6} = \frac{2 \pm 4}{6} = 1$ or $-\frac{1}{3}$.

Exercises 10.2

[1] $x^2 + 2x = 8 \Rightarrow x^2 + 2x + 1 = 8 + 1 \Rightarrow (x+1)^2 = 9 \Rightarrow x+1 = \pm 3 \Rightarrow$
$x = -1 \pm 3 \Rightarrow x = -4 \text{ or } 2$.

[5] $2y^2 = 16y - 38 \Rightarrow \frac{1}{2}(2y^2) = \frac{1}{2}(16y - 38) \Rightarrow y^2 = 8y - 17 \Rightarrow y^2 - 8y = -17 \Rightarrow$
$y^2 - 8y + 16 = -17 + 16 \Rightarrow (y-4)^2 = -1 \Rightarrow y - 4 = \pm\sqrt{-1} \Rightarrow$
$y - 4 = \pm i \Rightarrow y = 4 \pm i$.

[9] To solve $x^2 + 2x = 8$ using the quadratic formula, move the 8 to the left side: $x^2 + 2x - 8 = 0$.
Then $a = 1$, $b = 2$, and $c = -8$ and $x = \dfrac{-2 \pm \sqrt{4 - 4(1)(-8)}}{2} = \dfrac{-2 \pm \sqrt{4 + 32}}{2} =$
$\dfrac{-2 \pm \sqrt{36}}{2} = \dfrac{-2 \pm 6}{2}$. Therefore $x = -4$ or 2.

[13] If $4x^2 + 1 = 0$ then $a = 4$, $b = 0$, and $c = 1$. Then $x = \dfrac{0 \pm \sqrt{0 - 4(4)(1)}}{8} = \dfrac{\pm\sqrt{-16}}{8} =$
$\dfrac{\pm 4i}{8} = \pm\frac{1}{2}i$.

[17] $x^2 = 2x - 2 \Rightarrow x^2 - 2x + 1 = -2 + 1 \Rightarrow (x-1)^2 = -1 \Rightarrow x - 1 = \pm\sqrt{-1} \Rightarrow$
$x - 1 = \pm i \Rightarrow x = 1 \pm i$.

[21] $9v^2 + 25 = 0 \Rightarrow 9v^2 = -25 \Rightarrow v^2 = -\frac{25}{9} \Rightarrow v = \pm\sqrt{-\frac{25}{9}} \Rightarrow v = \pm\frac{5}{3}i$.

[25] $9k^2 + 1 = 3k$ is equivalent to $9k^2 - 3k + 1 = 0$. Therefore
$k = \dfrac{3 \pm \sqrt{9 - 4(9)(1)}}{18} = \dfrac{3 \pm \sqrt{-27}}{18} = \dfrac{3 \pm \sqrt{27}i}{18} = \dfrac{3 \pm \sqrt{9}\sqrt{3}i}{18} = \dfrac{3 \pm 3\sqrt{3}i}{18} =$
$\dfrac{3}{18} \pm \dfrac{3\sqrt{3}i}{18} = \dfrac{1}{6} \pm \dfrac{\sqrt{3}}{6}i$.

Exercises 10.2

[29] $2x^2 - 12x + 18 = 0 \Rightarrow x^2 - 6x + 9 = 0 \Rightarrow (x-3)^2 = 0 \Rightarrow x - 3 = \pm\sqrt{0} \Rightarrow x - 3 = 0 \Rightarrow x = 3$. Therefore 3 is the only solution.

[33] To solve $(x-6)(x-4) = -34$ we must 1st expand the left side and combine like terms. This gives $x^2 - 4x - 6x + 24 = -34 \Rightarrow x^2 - 10x + 58 = 0$. Then

$$x = \frac{10 \pm \sqrt{100 - 4 \cdot 1 \cdot 58}}{2} = \frac{10 \pm \sqrt{-132}}{2} = \frac{10 \pm i\sqrt{132}}{2} = \frac{10 \pm \sqrt{4 \cdot 33}\,i}{2} =$$

$$\frac{10 \pm 2\sqrt{33}\,i}{2} = \frac{10}{2} \pm \frac{2\sqrt{33}\,i}{2} = 5 \pm \sqrt{33}\,i.$$

[37] $-2x = \dfrac{1}{2x+1} \Rightarrow -2x \cdot (2x+1) = 1 \Rightarrow -4x^2 - 2x = 1 \Rightarrow -4x^2 - 2x - 1 = 0 \Rightarrow$

$4x^2 + 2x + 1 = 0 \Rightarrow x = \dfrac{-2 \pm \sqrt{4 - 4 \cdot 4 \cdot 1}}{8} = \dfrac{-2 \pm \sqrt{-12}}{8} = \dfrac{-2 \pm \sqrt{12}\,i}{8} =$

$\dfrac{-2 \pm 2\sqrt{3}\,i}{8} = \dfrac{-2}{8} \pm \dfrac{2\sqrt{3}\,i}{8} = -\dfrac{1}{4} \pm \dfrac{\sqrt{3}}{4}i.$

[41] To solve $\dfrac{y+1}{y-3} = 6 + \dfrac{4}{y-3}$, get rid of the fractions by multiplying both sides by $y - 3$:

$(y-3) \cdot \dfrac{y+1}{y-3} = (y-3) \cdot 6 + (y-3) \cdot \dfrac{4}{y-3} \Rightarrow y + 1 = 6y - 18 + 4 \Rightarrow 15 = 5y \Rightarrow y = 3$.

But $y = 3$ is not in the domain of the equation. Therefore there are no solutions

[45] The discriminant of $3x^2 - x + 4 = 0$ is $b^2 - 4ac = (-1)^2 - 4(3)(4) = 1 - 48 = -47$. Therefore the roots of the equation are imaginary numbers.

[49] Problem Statement: The sum of two numbers is 20 and their product is 64. Find the numbers.

Let x be the 1st number and $20 - x$ the 2nd number. Then $x(20-x) = 64 \Rightarrow$
$20x - x^2 = 64 \Rightarrow -x^2 + 20x - 64 = 0 \Rightarrow x^2 - 20x + 64 = 0 \Rightarrow$
$(x-4)(x-16) = 0 \Rightarrow x = 4$ or 16. If $x = 4$ then $20 - x = 16$, and if $x = 16$ then
$20 - x = 4$. In either case the two numbers are 4 and 16.

Exercises 10.2

53 Problem Statement: The sum of a number and its reciprocal is 2. Find the number.

Let x be the number. Then $\frac{1}{x}$ is the reciprocal and we know that $x + \frac{1}{x} = 2 \Rightarrow x \cdot \left(x + \frac{1}{x}\right) = x \cdot 2 \Rightarrow x^2 + 1 = 2x \Rightarrow x^2 - 2x + 1 = 0 \Rightarrow (x-1)^2 = 0 \Rightarrow x = 1$.
Therefore the number is 1.

57 If $2x(2x^2 + x + 8) = 0$ then $2x = 0$ or $2x^2 + x + 8 = 0$. Then $x = 0$ or

$$x = \frac{-1 \pm \sqrt{1 - 4(2)(8)}}{4} = \frac{-1 \pm \sqrt{-63}}{4} = -\frac{1}{4} \pm \frac{\sqrt{63}\,i}{4}$$. Therefore there are 3 solutions:

$x = 0$ and $x = -\frac{1}{4} \pm \frac{\sqrt{63}}{4}i$.

61 The equation $3x^2 + 2ix + 5 = 0$ can be solved by using the quadratic formula where $a = 3$, $b = 2i$, and $c = 5$.

Then $x = \dfrac{-2i \pm \sqrt{(2i)^2 - 4(3)(5)}}{6} = \dfrac{-2i \pm \sqrt{4i^2 - 60}}{6} = \dfrac{-2i \pm \sqrt{-4 - 60}}{6} =$

$\dfrac{-2i \pm \sqrt{-64}}{6} = \dfrac{-2i \pm 8i}{6} = \dfrac{-10i}{6}$ or $\dfrac{6i}{6}$. Therefore the solutions are $-\dfrac{5}{3}i$ and i.

65 A quadratic equation with real values of a, b, and c can not have one real root and one imaginary root. If the discriminant is positive there are 2 real roots; if the discrimant is 0, there is one real root; and if the discriminant is negative there are 2 imaginary roots.

Review Problems

69 The domain of $\dfrac{2}{x-4}$ is $\{x / x \neq 4\}$.

73 If $y = 2x + 1$ and $x = -1$ then $y = 2(-1) + 1 = -2 + 1 = -1$.

77 If $y = 2x^2 - x + 4$ and $x = 2$ then $y = 2 \cdot 2^2 - 2 + 4 = 2 \cdot 4 - 2 + 4 = 8 - 2 + 4 = 10$.

Exercises 10.3

[1] The set of ordered pairs $\{(1, 5), (2, 8), (7, 6)\}$ is a function because there are not two ordered pairs with the same 1st coordinate. The domain is the set of 1st components: $\{1, 2, 7\}$. The range is the set of 2nd components: $\{5, 6, 8\}$.

[5] The set of ordered pairs $\{(-3, 81), (-2, 16), (1, 1), (2, 16), (3, 81)\}$ is a function. Note that it is a function despite the fact that some of the ordered pairs have the same 2nd components. The domain is $(-3, -2, 1, 2, 3\}$ and the range is $\{1, 16, 81\}$.

[9] The equation $y = -2x - 1$ defines a function because each value of x produces just one value of y. The domain is $\Re$ because any real number can be replaced for x in the equation. The range is also $\Re$ as can be seen from the graph below:

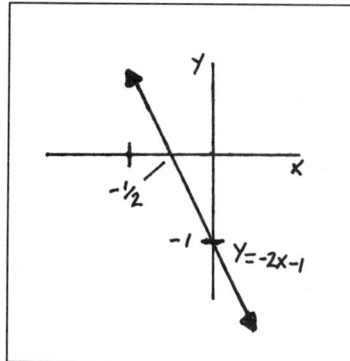

Figure 9

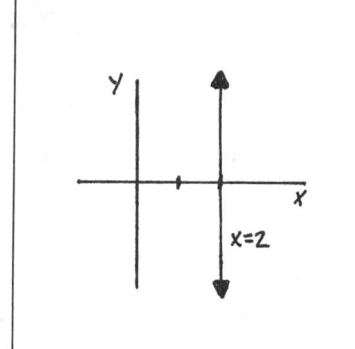

Figure 13

[13] The equation $x = 2$ does not define a function. To see this, examine the graph and note that there are an infinite number of points with x-coordinate 2.

[17] The domain of $y = \dfrac{3}{x+1}$ is the set of all x values where $x + 1 \neq 0 \Rightarrow x \neq -1$. Therefore the domain is $\{x/x \neq -1\}$.

[21] The domain of $y = \sqrt{1 - 3x}$ is the set of all x values where $1 - 3x \geq 0 \Rightarrow 1 \geq 3x \Rightarrow \frac{1}{3} \geq x \Rightarrow x \leq \frac{1}{3}$. Therefore the domain is $\{x/x \leq \frac{1}{3}\}$.

Exercises 10.3

25 The graph represents a function because every vertical line intersects the graph at no more than one point.

29 This graph does not represent a function because there are vertical lines which intersect the graph at more than one point. Note that the vertical line $x = 0$ intersects the graph at 5 points.

33 This graph is not a function because vertical lines intersect the graph at infinitely many points.

37 If $f(x) = \dfrac{x}{\sqrt{4x - x^2}}$ then

a) $f(-3) = \dfrac{-3}{\sqrt{4(-3)-(-3)^2}} = \dfrac{-3}{\sqrt{-12-9}} = \dfrac{-3}{\sqrt{-21}}$ which is not a real number.

Therefore -3 is not in the domain of the function.

b) $f(1) = \dfrac{1}{\sqrt{4 \cdot 1 - 1^2}} = \dfrac{1}{\sqrt{3}} = \dfrac{1}{\sqrt{3}} \cdot \dfrac{\sqrt{3}}{\sqrt{3}} = \dfrac{\sqrt{3}}{3}$.

c) $f(4) = \dfrac{4}{\sqrt{4 \cdot 4 - 4^2}} = \dfrac{4}{\sqrt{16-16}} = \dfrac{4}{\sqrt{0}} = \dfrac{4}{0}$ which is not a real number.

Note that $\dfrac{4}{0}$ is undefined and 4 is not in the domain of the function.

41 If $g(x) = \sqrt{x-1}$ then

a) $g(-2) = \sqrt{-2-1} = \sqrt{-3}$ which is not a real number.

b) $g(0) = \sqrt{0-1} = \sqrt{-1}$ which is not a real number.

c) $g(5) = \sqrt{5-1} = \sqrt{4} = 2$.

45 Since $g(x) = x^2 - 2x$, $-g(a) = -(a^2 - 2a) = -a^2 + 2a$.

49 To find $f(a-1)$ replace x by $a-1$. Since $f(x) = 3x$, $f(a-1) = 3(a-1) = 3a - 3$.

53 If $h(x) = \sqrt{9-x}$ then $h(9) = \sqrt{9-9} = 0$, $h(0) = \sqrt{9-0} = 3$, and $h(-7) = \sqrt{9-(-7)} = \sqrt{16} = 4$. Therefore the range is $\{0, 3, 4\}$.

Exercises 10.3

57 Given $x^2 + y^2 = 1$, note that $(0, 1)$ and $(0, -1)$ satisfy the equation. Therefore the equation does not define a function because 2 ordered pairs have the same x-coordinate.

Chapter Review Exercises

1. $\sqrt{-9} = i\sqrt{9} = 3i$.

5. $3x^2 = -24 \Rightarrow x^2 = -8 \Rightarrow x = \pm\sqrt{-8} = \pm i\sqrt{8} = i\sqrt{4 \cdot 2} = 2i\sqrt{2}$.

9. $(2 - 7i) - (6 - 5i) = 2 - 7i - 6 + 5i = (2 - 6) + (-7i + 5i) = -4 - 2i$.

13. $\dfrac{8i}{2+2i} = \dfrac{8i}{2+2i} \cdot \dfrac{2-2i}{2-2i} = \dfrac{16i - 16i^2}{4 - 4i^2} = \dfrac{16i - 16(-1)}{4 - 4(-1)} = \dfrac{16 + 16i}{4+4} = \dfrac{16}{8} + \dfrac{16i}{8} = 2 + 2i$.

17. $x^2 + 4x + 12 = 0 \Rightarrow x = \dfrac{-4 \pm \sqrt{16 - 4 \cdot 1 \cdot 12}}{2} = \dfrac{-4 \pm \sqrt{16 - 48}}{2} = \dfrac{-4 \pm \sqrt{-32}}{2} = \dfrac{-4 \pm i\sqrt{32}}{2} = \dfrac{-4 \pm 4\sqrt{2}i}{2} = -2 \pm 2\sqrt{2}i$.

21. $2y^2 + 13 = 2y \Rightarrow 2y^2 - 2y + 13 = 0 \Rightarrow y = \dfrac{2 \pm \sqrt{4 - 4 \cdot 2 \cdot 13}}{4} = \dfrac{2 \pm \sqrt{-100}}{4} = \dfrac{2 \pm 10i}{4} = \dfrac{1}{2} \pm \dfrac{5}{2}i$.

25. The set of ordered pairs $\{(1, 3), (2, 8), (3, 9)\}$ is a function. Its domain is $\{1, 2, 3\}$ and its range is $\{3, 8, 9\}$.

29. The domain of $y = \sqrt{3x + 9}$ is the set of x values where $3x + 9 \geq 0 \Rightarrow 3x \geq -9 \Rightarrow x \geq -3$. Therefore the domain is the set $\{x/x \geq -3\}$.

Chapter 10 Review Exercises and Test

33. If $f(x) = 2x + 4$ then $f(-6) = 2(-6) + 4 = -12 + 4 = -8$.

37. If $g(x) = x^2 + 2x + 3$ then $a + g(a) = a + (a^2 + 2a + 3) = a^2 + 3a + 3$.

Chapter Test

1. The graph of the complex numbers $\{-2, 5, i, -1 - 2i, 2 + 3i, 3 - 2i, -3 + 4i\}$ is given below.

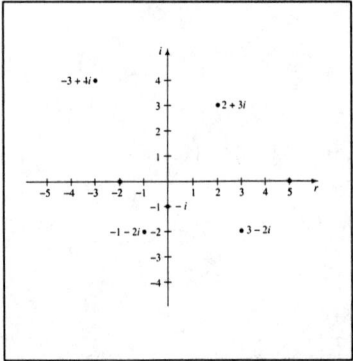

5. $\dfrac{i}{2-i} = \dfrac{i}{2-i} \cdot \dfrac{2+i}{2+i} = \dfrac{2i+i^2}{4-i^2} = \dfrac{-1+2i}{4-(-1)} = \dfrac{-1+2i}{5} = -\dfrac{1}{5} + \dfrac{2}{5}i$.

9. $x^2 + 2x = -5 \Rightarrow x^2 + 2x + 1 = -5 + 1 \Rightarrow (x+1)^2 = -4 \Rightarrow x + 1 = \pm\sqrt{-4} \Rightarrow x + 1 = \pm 2i \Rightarrow x = -1 \pm 2i$.

13. $3x^2 = 5x \Rightarrow 3x^2 - 5x = 0 \Rightarrow x(3x - 5) = 0 \Rightarrow x = 0$ or $3x - 5 = 0 \Rightarrow x = 0$ or $\dfrac{5}{3}$.

17. $f(x) = 3x - 4$ defines a function because each x produces just one value for y. The domain is all real numbers because any real number can be substituted for x. The range is all real numbers because $f(x)$ or $3x - 4$ can be any real number.

21. If $g(x) = 3x^2 + x$ then $g(-2) = 3 \cdot (-2)^2 + (-2) = 3 \cdot 4 - 2 = 12 - 2 = 10$.

25. Problem Statement: The sum of two numbers is 2 and their product is 2. Find the numbers.

Let x and $2-x$ be the numbers. Then $x(2-x) = 2 \Rightarrow 2x - x^2 = 2 \Rightarrow -x^2 + 2x = 2 \Rightarrow x^2 - 2x = -2 \Rightarrow x^2 - 2x + 1 = -2 + 1 \Rightarrow (x-1)^2 = -1 \Rightarrow x - 1 = \pm\sqrt{-1} \Rightarrow x = 1 \pm i$.

If $x = 1 + i$ then $2 - x = 2 - (1 + i) = 2 - 1 - i = 1 - i$.

If $x = 1 - i$ then $2 - x = 2 - (1 - i) = 2 - 1 + i = 1 + i$.

In either case the two numbers are $1 \pm i$.

Cumulative Review Exercises

1. $-6^2 + 20 = -36 + 20 = -16$

5. $\dfrac{4x - y^3}{5x + 3y} = \dfrac{4(3) - (-2)^3}{5(3) + 3(-2)} = \dfrac{12 - (-8)}{15 - 6} = \dfrac{12 + 8}{9} = \dfrac{20}{9}$.

9. The domain of $\dfrac{x-2}{x^2-9}$ is the set of x where $x^2 - 9 \neq 0$. $x^2 - 9 = 0 \Rightarrow x^2 = 9 \Rightarrow x = \pm 3$.
Therefore the domain is $\{x / x \neq \pm 3\}$.

13. We know $a^2 + b^2 = c^2$ where $a = 3$ and $c = 7$. Therefore
$3^2 + b^2 = 7^2 \Rightarrow 9 + b^2 = 49 \Rightarrow b^2 = 40 \Rightarrow b = \sqrt{40} = \sqrt{4 \times 10} = 2\sqrt{10}$ cm.

17. $(-2x)^3 = (-2)^2 x^3 = -8x^3$.

21. $\left(\dfrac{x^4}{6}\right)^2 = \dfrac{x^{4 \cdot 2}}{6^2} = \dfrac{x^8}{36}$.

25. $(x+3)(x^2 - 3x + 9) = x(x^2 - 3x + 9) + 3(x^2 - 3x + 9) = x^3 - 3x^2 + 9x + 3x^2 - 9x + 27 = x^3 + 27$.

29. $3x^3 - 6x^2 + 3x = 3x(x^2 - 2x + 1) = 3x(x-1)(x-1) = 3x(x-1)^2$.

33. To factor $9x^2 + 21xy + 4y^2$ find m and n such that $mn = 36$ and $m + n = 12$. Select $m = 6$ and $n = 6$. Then $9x^2 + 21xy + 4y^2 = 9x^2 + 6xy + 6xy + 4y^2 = 3x(3x + 2y) + 2y(3x + 2y) = (3x + 2y)(3x + 2y) = (3x + 2y)^2$.

37. $\dfrac{x^2 - 9}{a^2 - b^2} \div \dfrac{9 - 3x}{a + b} = \dfrac{(x+3)(x-3)}{(a+b)(a-b)} \div \dfrac{3(3-x)}{a+b} = \dfrac{(x+3)(x-3)}{(a+b)(a-b)} \cdot \dfrac{a+b}{3(3-x)} =$
$\dfrac{(x+3)(x-3)}{(a+b)(a-b)} \cdot \dfrac{-(a+b)}{3(x-3)} = \dfrac{-(x+3)}{3(a-b)}$.

Cumulative Review Exercises: Chapters 1 - 10

41. The graph of $y = -2$ is a horizontal line with y-intercept -2.

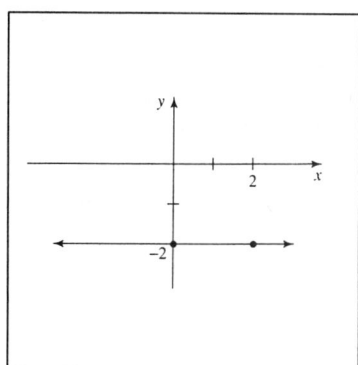

Figure 41

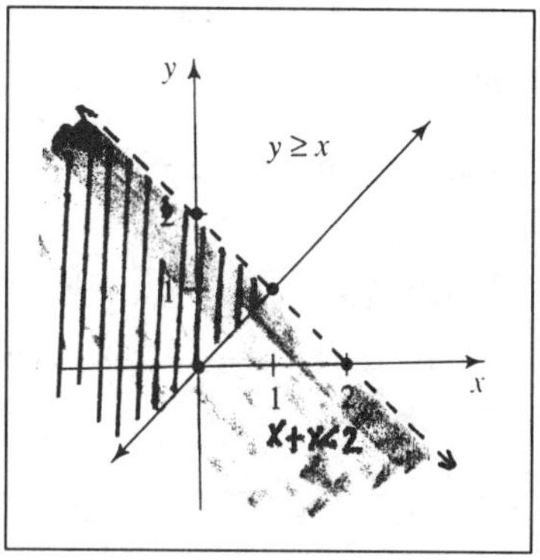

Figure 45

45. The graph of $\begin{array}{c} y \geq x \\ x + y < 2 \end{array}$ is the intersection of the 2 half-planes. $y \geq x$ is a half-plane bounded by, and including the line $y = x$. Since the point $(0, 1)$ satisfies the inequality $y \geq x$, the half-plane must open upward to include this point. $x + y < 2$ is the half-plane bounded by, but not including the line $x + y = 2$. This half-plane opens downward because $(0, 0)$ satisfies the inequality and the half-plane must include this point.

49. The line through $(4, 1)$ and $(4, 3)$ is a vertical line where each point has x-coordinate 4. Therefore the equation of the line is $x = 4$.

53. $\sqrt{-25} = i\sqrt{25} = 5i$.

57. $\sqrt[3]{-125} = \sqrt[3]{(-5)^3} = -5$.

61. $\dfrac{(1+\sqrt{2})(2+\sqrt{3})}{(2-\sqrt{3})(2+\sqrt{3})} = \dfrac{2+\sqrt{3}+2\sqrt{2}+\sqrt{6}}{4-3} = 2+\sqrt{3}+2\sqrt{2}+\sqrt{6}$.

65. $s = \sqrt{\dfrac{t}{5}} \Rightarrow s^2 = \left(\sqrt{\dfrac{t}{5}}\right)^2 \Rightarrow s^2 = \dfrac{t}{5} \Rightarrow t = 5s^2$.

69. $8x - 1 > x + 20 \Rightarrow 7x > 21 \Rightarrow x > 3$.

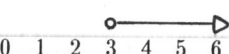

73. $x^3 - 25x = 0 \Rightarrow x(x^2 - 25) = 0 \Rightarrow x(x+5)(x-5) = 0 \Rightarrow x = 0, x+5 = 0$, or $x - 5 = 0$. Therefore $x = 0, 5$, or -5.

77. If $2x^2 - 3x - 1 = 0$ then $x = \dfrac{3 \pm \sqrt{3^2 - 4(2)(-1)}}{4} = \dfrac{3 \pm \sqrt{9+8}}{4} = \dfrac{3 \pm \sqrt{17}}{4}$.

81. Let x be the larger number and y the smaller number. Then $x + y = 20$ and $x - 10y = 75$.
Add -1 times the 1st equation to the 2nd equation:
$$\begin{array}{r} -x - y = -20 \\ x - 10y = 75 \\ \hline -11y = 55 \end{array}$$ $\Rightarrow y = -5$ and $x + (-5) = 20 \Rightarrow x = 25$.
Therefore the larger number is 25 and the smaller number is -5.

85. Let x be the width and $x + 2$ the length. Since the perimeter 28 cm,
$2x + 2(x+2) = 28 \Rightarrow 2x + 2x + 4 = 28 \Rightarrow 4x = 24 \Rightarrow x = 6$.
Therefore the width is 6 cm and the length is 8 cm.

89. Let x be the time at 40 mph and y the time at 52 mph. Then $x + y = 5$ and since $d = rt$, $40x + 52y = 224$. Add -40 times the 1st equation to the 2nd equation:

$$\begin{array}{r} -40x - 40y = -200 \\ 40x + 52y = 224 \\ \hline 12y = 24 \end{array}$$ $\Rightarrow y = 2$ and $x = 3$.

Then the time at 40 mph is 3 hours and the time at 52 mph is 2 hours.
Therefore the distance traveled at 40 mph is $3 \times 40 = 120$ miles and the distance traveled at 52 mph is $2 \times 52 = 104$ miles.

93. Let x be the number of nickels and y the number of dimes. Then
$x + y = 18$ and $5x + 10y = 145$. From the 1st equation we see that $y = 18 - x$. Then
$5x + 10(18 - x) = 145 \Rightarrow 5x + 180 - 10x = 145 \Rightarrow 35 = 5x \Rightarrow x = 7$ and $y = 18 - 7 = 11$.
Therefore there are 7 nickels and 11 dimes.

97. Let x be the amount of 6% solution and y the amount of 15% solution.
Then $x + y = 3$ and $.06x + .15y = .12(3)$. Multiply the 2nd equation by 100 which gives $6x + 15y = 36$. Adding -6 times the 1st equation to this last equation gives

$$\begin{array}{r} -6x - 6y = -18 \\ 6x + 15y = 36 \\ \hline 9y = 18 \end{array} \quad \Rightarrow \quad y = 2 \text{ and } x = 1.$$

Therefore mix 1 liter of 6% solution and 2 liters of 15% solution.

Appendix

Exercise Set A.1

1. Since 2 is an element of the set $A = \{0, 1, 2, 3\}$, $2 \in A$.

5. $\{0, 1, 2, 3, 4, 5\}$ is the set consisting of the whole numbers from 0 to 5. Therefore $\{0, 1, 2, 3, 4, 5\} = \{x \in W/\ 0 \leq x \leq 5\}$.

9. $\{5, 6, 7, \ldots\}$ is the set of whole numbers (or natural numbers) greater than or equal to 5. Therefore $\{5, 6, 7, \ldots\} = \{x \in W/\ x \geq 5\} = \{x \in N/\ x \geq 5\}$.

13. The set $\{x \in Z/x < 7\}$ is the set of integers less than 7. Therefore $\{x \in Z/\ x < 7\} = \{6, 5, 4, 3, 2, 1, 0, -1, -2, \ldots\}$.

17. B is a subset of A because each element of B is also an element of A. Therefore $B \subseteq A$.

21. Since every element of C is an element of C, C is a subset of C. Therefore $C \subseteq C$.

25. The subsets of $\{1, 2\}$ are $\{1\}, \{2\}, \{1, 2\}$, and $\emptyset$.

29. If $A = \{1, 3, 5, 7\}$ and $B = \{1, 2, 3, 4\}$, $A \cup B = \{1, 2, 3, 4, 5, 7\}$.

33. $A \cap C = \{1, 3, 5, 7\} \cap \{12, 13\} = \emptyset$ because the sets have no elements in common.

37. If we combine the rational numbers with the irrational numbers we get the set of real numbers. Therefore $Q \cup H = \Re$.

41. Since a natural number is a rational number, the natural numbers and the irrational numbers have nothing in common. Therefore $N \cap H = \emptyset$.

45. If $A = \{1, 2, 3, 4\}$ and $B = \{3, 4, 5, 6\}$ then $A \cup B = \{1, 2, 3, 4, 5, 6\}$.

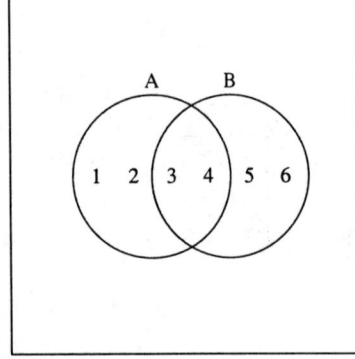

Figure 45

Figure 49

Appendix

49. $C \cup A = \{1, 2\} \cup \{1, 2, 3, 4\} = \{1, 2, 3, 4\}$.

53. If we combine the elements of A with the empty set we get no additional elements. Therefore $\emptyset \cup A = A$.

57. If $A \subset B$ and $B \subset C$ then A must also be a subset of C. That is $A \subseteq C$.

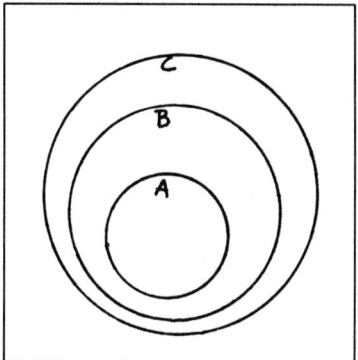

Figure 57

Figure 61

61. We know $A \subseteq C$ (see problem 57) . Therefore every element of A is an element of C and $A \cap C = A$.

65. Since A is contained in B and B is contained in C, the intersection of the 3 sets is A . Therefore $A \cap B \cap C = A$.

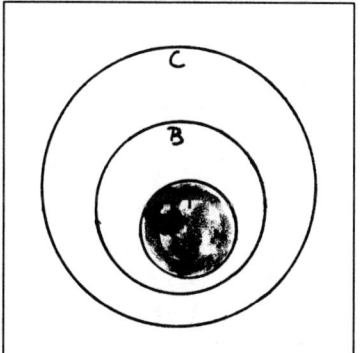

Appendix

Exercise Set B.1

1. The fraction $\frac{5}{6}$ has numerator 5 and denominator 6.

5. $1\frac{3}{4}$ is a mixed number.

9. 17 is prime.

13. $54 = 2 \times 27 = 2 \times 3 \times 3 \times 3 = 2 \times 3^3$.

17. $144 = 12 \times 12 = 2 \times 2 \times 3 \times 2 \times 2 \times 3 = (2 \times 2 \times 2 \times 2)(3 \times 3) = 2^4 \times 3^2$.

21. $\frac{75}{27} = \frac{25 \times 3}{9 \times 3} = \frac{25}{9}$.

25. $\frac{48}{32} = \frac{16 \times 3}{16 \times 2} = \frac{3}{2}$.

29. $\frac{18}{144} = \frac{18 \times 1}{18 \times 8} = \frac{1}{8}$.

33. $\frac{2}{3} \div \frac{5}{7} = \frac{2}{3} \times \frac{7}{5} = \frac{2 \times 7}{3 \times 5} = \frac{14}{15}$.

37. $\frac{25}{12} \div \frac{75}{32} = \frac{25}{12} \times \frac{32}{75} = \frac{25 \times 32}{12 \times 75} = \frac{25 \times 4 \times 8}{25 \times 3 \times 4 \times 3} = \frac{8}{9}$.

41. $\frac{1}{9} \div \frac{14}{18} = \frac{1}{9} \div \frac{2 \times 7}{2 \times 9} = \frac{1}{9} \div \frac{7}{9} = \frac{1}{9} \times \frac{9}{7} = \frac{1 \times 9}{7 \times 9} = \frac{1}{7}$.

45. $\frac{7}{8} - \frac{3}{8} = \frac{7-3}{8} = \frac{4}{8} = \frac{1}{2}$.

49. The LCD of $\frac{7}{15}$ and $\frac{3}{20}$ is 60. Therefore $\frac{7}{15} - \frac{3}{20} = \frac{7}{15} \cdot \frac{4}{4} - \frac{3}{20} \cdot \frac{3}{3} = \frac{28}{60} - \frac{9}{60} = \frac{28-9}{60} = \frac{19}{60}$.

53. $\frac{3}{2} + \frac{7}{10} = \frac{3}{2} \cdot \frac{5}{5} + \frac{7}{10} = \frac{15}{10} + \frac{7}{10} = \frac{22}{10} = \frac{2 \times 11}{2 \times 5} = \frac{11}{5}$.

57. $1\frac{5}{6} = 1 + \frac{5}{6}$. Therefore $1\frac{5}{6} + \frac{1}{6} = 1 + \frac{5}{6} + \frac{1}{6} = 1 + \frac{6}{6} = 1 + 1 = 2$.

61. $\frac{7}{8} \div \frac{14}{16} = \frac{7}{8} \div \frac{7 \times 2}{8 \times 2} = \frac{7}{8} \div \frac{7}{8} = \frac{7}{8} \times \frac{8}{7} = \frac{56}{56} = 1$.

65. The LCD is 12. Then $\frac{5}{12} + \frac{1}{4} + \frac{1}{6} = \frac{5}{12} + \frac{1}{4} \cdot \frac{3}{3} + \frac{1}{6} \cdot \frac{2}{2} = \frac{5}{12} + \frac{3}{12} + \frac{2}{12} = \frac{5+3+2}{12} = \frac{10}{12} = \frac{5 \times 2}{6 \times 2} = \frac{5}{6}$.

Appendix

69. To find the LCD, factor each denominator into a product of prime numbers:

$22 = 2 \times 11$

$99 = 3^2 \times 11$

$66 = 2 \times 3 \times 11$.

The LCD is $2^a 3^b 11^c$ where a is the highest power on 2, b is the highest power on 3, and c is the highest power on 11. Thus we see $a = 1$, $b = 2$, and $c = 1$ and the LCD is

$2^1 \times 3^2 \times 11 = 2 \times 9 \times 11 = 198$.

So $\frac{3}{22} + \frac{2}{99} + \frac{5}{66} = \frac{3}{22} \cdot \frac{9}{9} + \frac{2}{99} \cdot \frac{2}{2} + \frac{5}{66} \cdot \frac{3}{3} = \frac{27}{198} + \frac{4}{198} + \frac{15}{198} = \frac{27 + 4 + 15}{198} = \frac{46}{198} = \frac{23}{99}$.

73. The LCD is 24. Then $\frac{1}{8} + \frac{1}{6} + \frac{1}{4} = \frac{1}{8} \cdot \frac{3}{3} + \frac{1}{6} \cdot \frac{4}{4} + \frac{1}{4} \cdot \frac{6}{6} = \frac{3}{24} + \frac{4}{24} + \frac{6}{24} = \frac{13}{24}$.

Therefore she paid $\frac{13}{24}$ of her Mastercard bill. Since $1 - \frac{13}{24} = \frac{24}{24} - \frac{13}{24} = \frac{11}{24}$, she still owes $\frac{11}{24}$ of her bill.